Bumbreath Botox and Bubbles

Bumbreath Botox and Bubbles

and other **Fully Sick** Science Moments

Dr Karl Kruszelnicki

New Moments in Science #5

Illustrated by Adam Yazxhi

HarperCollins*Publishers*

HarperCollins*Publishers*

First published in Australia in 2003
by HarperCollins*Publishers* Pty Limited
ABN 36 009 913 517
A member of the HarperCollins*Publishers* (Australia) Pty Limited Group
www.harpercollins.com.au

HarperCollins*Publishers*
25 Ryde Road, Pymble, Sydney NSW 2073, Australia
31 View Road, Glenfield, Auckland 10, New Zealand
77–85 Fulham Palace Road, London W6 8JB, United Kingdom
2 Bloor Street East, 20th floor, Toronto, Ontario M4W 1A8, Canada
10 East 53rd Street, New York NY 10022, USA

National Library of Australia Cataloguing-in-publication data:

Kruszelnicki, Karl, 1948– .
Bumbreath, botox and bubbles.
ISBN 0 7322 6715 3.
1. Science – Popular works. 2. Discoveries in science.
3. Curiosities and wonders. I. Yazxhi, Adam. II. Title.
(Series: New moments in science; no. 5).
500

Cover photography: Jonathan Clabburn
Cover and internal design: Maxco Creative Media
Internal layout: Judi Rowe, Agave Creative Design
Illustrations: Adam Yazxhi
Printed and bound in Australia by Griffin Press on 80gsm Econoprint

6 5 4 3 03 04 05 06

for my family

contents

When she was about five years old, my little daughter Alice bounced into the bedroom one morning to give me a big hug and then immediately said, '*Hey Dad, you smell like a bum*.' She was right. Like most people, I suffered the dreaded 'Morning Breath'.

bumbreath

So I went looking in the medical and scientific literature, and found nothing. But in 1999, I was thrilled when I discovered that Dr Michael Levitt and his colleagues presented a preliminary report on my problem of Morning Breath at the Fourth International Conference on Breath Odor in Los Angeles. I found out how to fix my Morning Breath, and also discovered that bad breath comes from that moist bacterial jungle in your mouth.

Some of these mouth bacteria make truly nasty chemicals, some with disgusting names. There's 'putrescine' which gives decaying meat its special fragrance; 'cadaverine' which is usually given off by corpses; 'skatole' and 'methyl mercaptan' which smell like faeces; 'isovaleric acid' which smells like sweaty feet; and 'hydrogen sulphide' which smells like rotten eggs. The bacteria also emit 'indole', which smells lovely at low concentrations and is used sparingly in perfumes, but strangely smells foul at high levels. It shows how little we know about 'smells', that we don't understand why indole behaves likes this.

the problem

Many people think they suffer from bad breath.

In the USA, half a million of these people turn up to dentists each week complaining about their bad breath. The polite medical word is 'halitosis' — Latin *halitus* means 'breath', and the Greek -*osis* means 'abnormal condition of'. In fact, about 20% of the US population think they have a long-term problem with halitosis. To fix this problem, in the year 2000, Americans spent US$1.8 billion on toothpaste, US$950 million on toothbrushes and dental floss, US$740 million on mouthwashes and other dental rinses, US$715 million on oral-care gum and US$625 million on breath fresheners (excluding mouthwash and gum).

'Halitophobe' is the name given to people who worry so much about their bad breath that it interferes with their normal life.

Back in the old days, people wrongly believed that rotten food in your gut gave off bad smells which came up via your oesophagus into your mouth, and finally into somebody's face. But thanks to the work of Breath Scientists like Professor Mel Rosenberg, we now know better.

Mel Rosenberg is Professor of Microbiology at the Department of Oral Biology in Dental Medicine at the Tel-Aviv University School of Dental Medicine. He says most bad breath occurs in the mouth, and hardly any comes from below the tonsils.

Mel Rosenberg says people who are most at risk of bad breath (once all the other causes have been excluded) are people who talk a lot.

mel's meteoric rise

Alice Shirrell Kaswell and Stephen Drew honoured Mel Rosenberg with their article in the *Annals of Improbable Research*. They write of his many university and hospital positions around the world. They also quote some of his research papers that *'tell the exciting story of how one man helped a tiny field of research blossom, grow, and finally attain recognition as a distinct branch of science.'*

It didn't come easy — he had to work hard to get to his present pinnacle of fame. He says: *'Over the past 15 years I have had the opportunity to smell the mouths of thousands of individuals in clinics and in research studies, not to mention the hundreds I have smelled surreptitiously in supermarkets, in airplanes, and in synagogue during the fast day of Yom Kippur.'*

In the early years he wrote *'Bad Breath: Diagnosis And Treatment'* (1990) which was then followed by *'Measurement of Oral Malodor: Current Methods and Future Prospects'* (1992). He achieved international recognition in 1994 with his paper *'Conference Report: The First International Workshop on Oral Malodor'*.

Mel then showed his potential for creating rapid technical progress in the field of bad breath with his 1995 paper *'Self-estimation of Oral Malodor'* followed by his 1996 paper, *'Developments in Diagnosis and Treatment of Bad Breath'*.

In his psychological years, he wrote *'Self-Perception Of Breath Odor: Role Of Body Image And Psychopathologic Traits'* (2000).

In 2001, Mel's chosen field blossomed. The growing International Society for Breath Odor Research (ISBOR) had some 350 delegates attending its fifth international meeting in Tokyo.

He finally achieved his deserved recognition for having helped create a mature, freestanding branch of science with his article in the *Scientific American* called *'The Science of Bad Breath'* (2002).

Good Breath

Mel Rosenberg came up with seven tips for good breath.

1. Clean the back of your tongue with a scraper or a toothbrush.
2. Eat breakfast. It gets the saliva flowing and cleans the tongue.
3. Chew gum or drink water to stop your mouth from drying out.
4. Rinse and gargle with a mouthwash just before going to sleep. This slows the growth of bacteria overnight.
5. Clean your mouth after you eat/drink smelly foods/drinks such as garlic, curry or coffee. If you eat or drink foods high in proteins, clean between your teeth.
6. Of course, always floss and brush.
7. The most reliable way to check for bad breath is to ask a family member to smell you.

fish-odour syndrome and others (1%)

There are a few dozen obscure causes that make up about 1% of cases of bad breath.

An interesting and rare cause is *Trimethylaminuria*. It's also called 'Fish-Odour Syndrome', and there are well over 200 cases on record. Probably the first record of it is in the Indian *Mahabharata* (which in Sanskrit means 'Great Epic of the Bharata Dynasty'). The *Mahabharata* is one of the two major Sanskrit epics (the other is the *Ramayana*), which are great literary and religious works, and which deal with the evolution of Hinduism between 400 BC and 200 AD. The *Mahabharata* is huge — about seven times longer than the *Odyssey* and the *Iliad* combined. A small part of it tells the story of a young woman, Satyavata, who stank of 'rotting fish'. She was expelled from society because of her smell, but a demi-god took pity on Satyavata and bestowed a miracle cure on her.

But animals can get 'fish odour' too. A certain breed of Rhode Island Red chicken produces eggs which smell 'fishy', if the diet is rich in chemicals which lead to trimethylamine (formula $N(CH_3)_3$).

We all produce trimethylamine (which smells like bad fish) from the food we eat. Practically all of us make the enzyme that gets rid of trimethylamine. A very small number of people have low levels of this enzyme. So they have high levels of trimethylamine throughout their body. These molecules get picked up by the blood stream, travel into the lungs, jump into the airstream, and finally come out through the mouth — giving these poor unfortunates the persistent odour of rotting fish. The trimethylamine also comes out in the sweat and the urine (hence the name, trimethylaminuria). This condition can sometimes have a very severe impact on the sufferers, such as severe mental depression and rarely, attempted suicide.

cause of bad breath (3%)

In about 3% of cases, the tonsils cause bad breath.

The tonsils are part of your immune system. They have tiny little pits on the surface. Occasionally, bacteria live in these pits, and then get killed by your immune system. These conglomerations of dead bacteria and dead immune system cells usually are squirted out of the little pits, and then swallowed into your stomach.

But sometimes, these dead bacteria don't get squirted out, they stay in the pits and calcify. They get turned into tonsilloliths (tonsil stones) that are about half the size of a head of a match, which can smell quite nasty. (You might see one if you sneeze into a tissue.) They cause no other medical problems, and so many doctors and dentists have not heard of them.

inside the nose (5%)

In 5–10% of cases, bad breath is made in the nose — from sinusitis, or a reduced mucus flow, or rarely, a foreign body.

There was the curious case of a 28-year-old woman who went to her doctor complaining of persistent bad breath. When he looked up her nose, he found a small bead that she put up there as a young child. When he removed the foreign body, the bad breath went away — after a quarter of a century.

The quality of the bad breath that originates in the nose is quite different from the bad breath that comes from the mouth, and this can be a diagnostic tool.

back of the tongue (85–90%)

In 85–90% of cases, bad breath is generated in the mouth — practically all on the back of your tongue. (A small percentage of mouth-generated bad breath comes from diseased teeth or gums. A famous encyclopaedia wrongly claims that *'Halitosis is due to the rotting debris in the pockets under the gum margins.'*)

The tongue is triply famous. First, it's the only muscle in your body that gets stronger with age. Second, it's probably the muscle that's the most fun to exercise. And third, the tongue is like a shag-pile carpet. The bacteria responsible for bad breath live in the crypts (tiny holes) on the shag-pile surface of the tongue.

On average, you generate about 800–1500 ml of saliva each day. Saliva is good. First, it physically washes away nasty bacteria in your mouth, as well as the leftover food they like to eat. Second, saliva has various chemicals that help fight bacteria. And third, saliva contains various protein antibodies that kill some of the bacteria in your mouth.

When you go to sleep, two things happen to your lovely saliva. First, the production of saliva plummets. Second, many people breathe through their mouth, which dries out what little saliva there is.

The very back of your tongue is poorly washed by saliva. Food collects there, and the bacteria love to eat it. So every night, the bacteria multiply like crazy, and they feed on leftover food and dead mouth cells. They also feed on postnasal drip, which happens to about one in four people.

bumbreath

There is a difference between 'morning breath' (Bumbreath), which is at its worst when you wake up, and 'halitosis', which is there all the time. But they both have one thing in common — the bad odours seem to result from certain sulphur-containing gases in your breath.

In the Land of Science, everything has to be defined, and so Dr Levitt defines Morning Breath as *'malodorous breath upon wakening after a night's sleep'*.

Babe . . . you leave me breathless

bumbreath experiment

Levitt and his colleagues from the Minneapolis Veterans Affairs Medical Center and the Department of Medicine and the Department of Food Science and Nutrition at the University of Minnesota did a study with eight healthy adults aged between 27 and 51. 'Eight' is not really a large enough number to give you good statistics, but at least it's a start. The adults were all free of any dental disease, and none of them thought they had any halitosis.

The scientists collected gas samples from the mouth at various times during the day after waking up. These 'bad breath' gases almost qualify as Weapons of Mass Destruction. The scientists had to use polypropylene tubes and apparatus, because these 'bad-breath', sulphur-containing gases react with rubber and with glass! They also tried various treatments, in an effort to find what could reduce Morning Breath. They found three main sulphur-containing gases present in the samples of Morning Breath.

Hydrogen sulphide (H_2S), which smells like rotten eggs, had the highest concentration in Morning Breath. Its level was about 1.3 times greater than another sulphur-containing gas, methanethiol (CH_3SH), which smells like rotting cabbage.

And they also measured very small quantities of the gas, dimethylsulphide (CH_3SCH_3), which smells a little sweet. (Isn't that typical — the sweet-smelling gas is present in the lowest levels, about 16 times lower than the foul-smelling hydrogen sulphide. Scientists call this 'The Law of Maximum Inconvenience'.)

bumbreath removal

If the experimenters did nothing at all, the hydrogen sulphide in Bumbreath would decline significantly over the first hour after the subjects awoke. Then it would increase over the next seven hours back to its original full-flavoured, stinky level.

Brushing your teeth with toothpaste for two minutes is good for the health of your teeth but it did nothing for Bumbreath. Swallowing a few herbal breath-freshening capsules also had no effect. Mechanical scrubbing of the tongue (by food or brushing) reduced the Bumbreath gases.

Eating food, like breakfast, reduced the concentration of hydrogen sulphide (rotten eggs) and methanethiol (rotten cabbage). In fact, a dry hard bread roll was quite effective in reducing the levels.

Brushing the tongue (not the teeth) with a toothbrush for one minute with water also reduced these nasty gases. The trick is that you really have to scrub the very back of your tongue. The problem is that you will probably involuntarily gag (this is a protective reflex, to stop you from swallowing insects and other unwanted intruders into your mouth). The solution is to hum while reaching right down to the back of the tongue (I discovered this by watching nature movies in various hotel rooms — at least, I think they were nature movies). On the other hand, cleaning the tongue has been practised in the East for thousands of years.

The most effective treatment of all was to gargle and rinse the mouth with 5 ml of 3% hydrogen peroxide for one minute. This significantly reduced the concentration of all three gases, and kept them low for the next eight hours.

So if you want to reduce Bumbreath when you wake up, hydrogen peroxide is best, but simply brushing your tongue comes a close second.

who has halitosis?

As I mentioned earlier, people who suffer most from halitosis are those whose job involves talking a lot — for example, politicians, lawyers, teachers and radio science broadcasters. Their mouths dry out as they talk, so the half a millilitre of saliva they generate every minute gets absorbed by the dry mouth — and can't do its normal job of carrying away the bacteria.

Different nationalities around the world have different 'cures' for bad breath. Most Westerners believe mint can freshen your mouth. In Thailand, they chew the peels of oversized guavas, while Iraqis keep cloves between their teeth. Indians like to chew on fennel seeds, while Italians prefer parsley.

delusional halitosis

There's 'hunger breath' (bad breath from skipping meals), 'PMS breath' (the breath associated with a woman about to get her period) — and there's 'delusional halitosis', the bad breath you're having when you don't have bad breath.

Psychiatrists classify delusional halitosis as one of the 'monosymptomatic hypochondriacal psychoses'. 'Mono-symptomatic' means people who have a single symptom (in this case, thinking they have bad breath) in their particular psychosis.

'Hypochondriacal' literally means 'under the chest'. People who have no real pathology, but continually suffer from poorly-defined pain, will often point to the belly (which is under or below the chest) as the site of the 'pain'. It now refers to people who have nothing physically wrong with them.

'Psychosis' means an impairment of your thinking, so that you lose contact with reality (the reality in this case, is that your breath is perfectly normal).

In 'delusional halitosis', the person is deeply convinced their breath is really rank. The condition can be serious — sufferers have been known to go to such extreme measures such as not leaving their house, giving up their careers or attempting suicide. One person tried so hard to purify their breath, they used 100 ml of toothpaste every four hours. Another student sufferer would get into their class extra early, sit all alone in a distant corner, and then use the fact that all

Bumbreath . . . who is responsible?

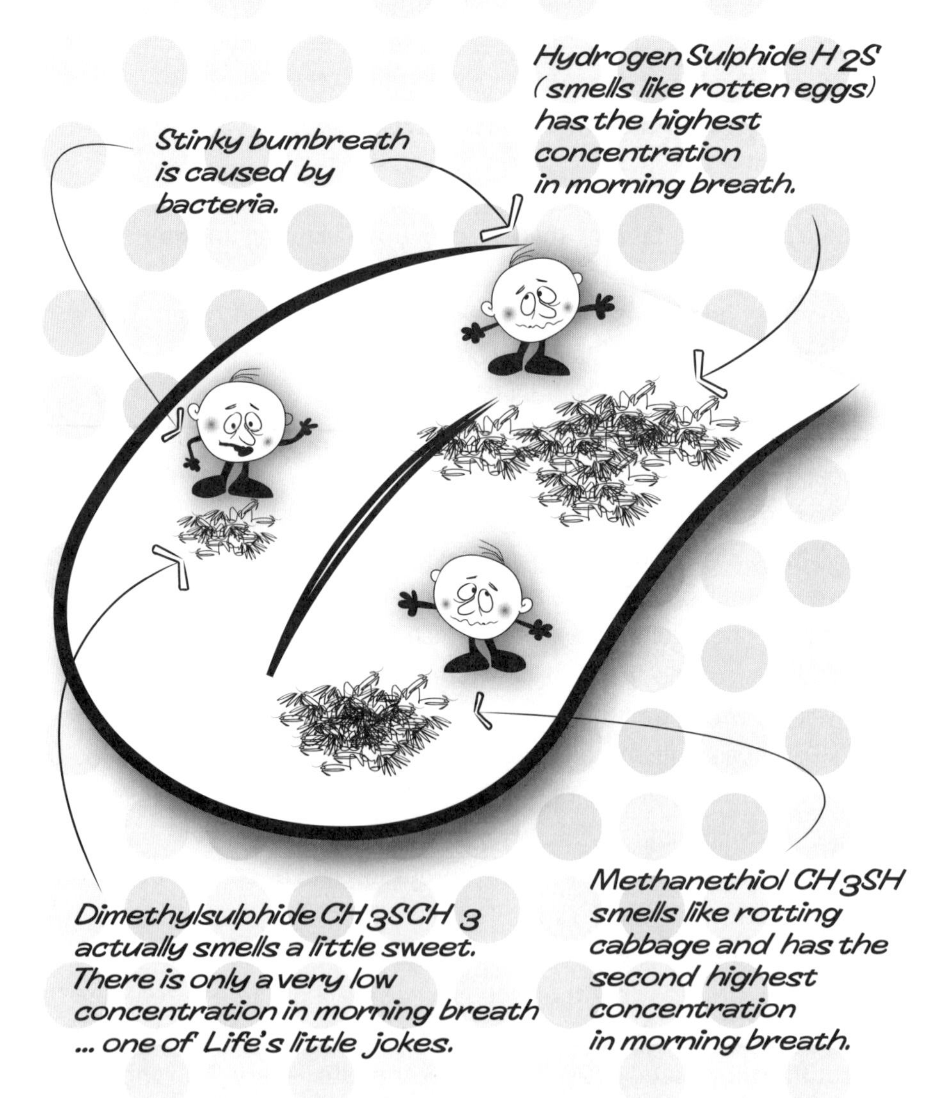

the other students sat closer to the front as proof that they had really bad breath.

One study looked at 32 people who all suffered delusional halitosis — and not one of them had any hint of bad breath.

Other 'monosymptomatic hypochondriacal psychoses' include the delusion you have small seed-like objects under the skin, the delusion your skin is infested by parasites, and the delusion your dental bite is abnormal (Phantom Bite Syndrome).

Halitosis and Snoring in a Dog

The owner of a 14-year-old crossbred dog brought his beloved pet, which had very bad breath and noisy breathing, to the Veterinary Clinic at the University of Sydney. The dog had vomited after eating a large meal of raw chicken, and then had developed very bad breath, a cough and a 'snoring' noise when it breathed.

The vets took x-rays of the head (where the noise was coming from) and the chest (because of the history of coughing). They saw 'a cylindrical radio-opaque foreign body' — what you and I would call a chicken bone.

The vet thought when the dog vomited up the chicken bones from its meal, one of them had impacted hard at the back of the nose and stayed there.

The immune system attacked this foreign body. This caused a local inflammation, which led to an accumulation of mucous and exudates, an awful lot of 'associated putrefaction' and a really foul odour.

The wedged chicken bone caused abnormal dripping of a nasty-smelling liquid (postnasal drip). When this dripped into the dog's airways, it made the dog cough.

The chicken bone was in the path of the air when the dog breathed in through its nose. The bone produced turbulent airflow, and this caused the snoring. That's how the bone caused the smell, the cough and the snore.

The treatment was easy. They anaesthetised the dog and *'with some difficulty it was possible to dislodge the bone by a gentle side-to-side rocking action and to remove it through the mouth'*. They recovered *'a chicken bone covered by foul-smelling tissue debris'*.

The vets at the hospital had previously found other foreign bodies at the back of the nose — not just bones, but tablets and even blades of grass. Again, they had most probably been rammed into position by very forceful vomiting.

smell own breath

How can you tell if your breath stinks? Think of the advantages if one could smell one's own bad breath. You could avoid the massive embarrassment of trying to discover from a friend if your own breath ponged, or of finding a delicate way to tell a friend that their breath reeked.

Luckily, in 1995, Mel Rosenberg looked at (and solved) this problem when he ran an 'Oral Malodor Clinic' in Tel-Aviv. He set up a study to see if people were '*able to smell their own malodor*'. Mel worked with 52 Israeli citizens, 43 of whom claimed that they had bad breath.

He had them run through some 'well-tried-and-possibly-true' techniques for inhaling and checking your own bad breath. He got them to cup their hands over their mouth and nose, and then breathe out through the mouth and in through the nose. He also got them to do the 'breathe–rebreathe' thing under the blankets, as well as smelling their freshly-used telephone mouth piece and their freshly-used piece of dental floss. He even got them licking their wrist '*in a perpendicular fashion*', and then sniffing it.

He then experimented with his own brand new technique. They were to spit into a laboratory dish that was then closed and cooked at 37°C for five minutes, and finally 'presented for odour assessment' to the spitter.

Of course, Dr Rosenberg would be the impartial 'odour judge'. This meant he had to sniff 52 mouths breathing upon him, 52 saliva-covered wrists, 52 telephone mouth pieces, 52 strands of dental floss, and 52 warmed-up samples of saliva in a laboratory dish. To make sure his nose didn't get overloaded and stop functioning normally, he would regularly sniff his control solution — '*a chicken-dung based fertiliser in aqueous suspension ... in an opaque sniff bottle*'.

The results were clear — there is no simple way to check if you have bad breath. There was absolutely no relationship between how the volunteers, and Rosenberg, rated people's breath odour.

The only test with any reliability was the 'incubated spit technique'. It's not simple, but you can monitor your own breath status — though you have to go to the trouble of carrying around a sealable laboratory dish, and 'cooking it' under your armpit or next to your body for five minutes. Or ask a friend.

Listerine: The Name

Listerine was named after Sir James Lister. Back in the 1860s, the only people who knew about bacteria were a few scientists with microscopes. Most doctors, if they had ever heard of them, thought bacteria had nothing to do with disease. James Lister campaigned against the lack of hygiene practised by most surgeons. Back then, surgeons didn't wash their hands, didn't sterilise bandages, and allowed spectators to crowd in close to the operating table. The death rate after surgery was as high as 90% — mostly because of infection. Eventually, the surgeons took up Lister's call for 'antiseptic surgery'. In St Louis, Dr Joseph Lawrence was so impressed that in 1880, he named his new anti-bacterial liquid after Lister — as Listerine.

References

Rosenberg, M., 'Bad breath: diagnosis and treament', *University of Toronto Dental Journal,* vol. 3, 1990, pp. 7–11.

Guyton, Arthur C. & Hall, John, *Text Book of Medical Physiology,* W.B. Saunders Company, 1996, pp. 815–819.

Laurendet, H., Govendir, M., Porges, W.L. & Malik, R., 'Snoring and halitosis in a dog', *Australian Vet. Journal,* vol. 76, no. 4, April 1998, pp. 245, 250–251.

Levitt, M.D., *et al.*, 'Morning breath odor: influence of treatments on sulfur gases', *Journal of Dental Research,* vol. 79, no. 10, July 2000, pp. 1773–1777.

'The science of bad breath', *Scientific American*, April 2002, pp. 58–65.

They say nobody knows what the nose knows — but now scientists have begun to work out just what the nose does know. And it knows sex ...

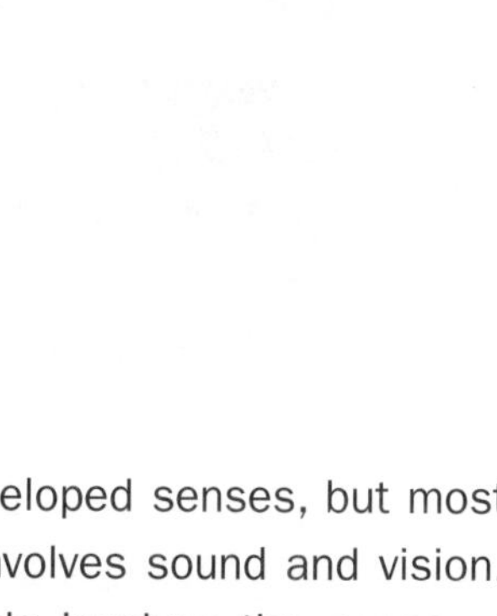

sex organ up your nose

We humans have a number of highly-developed senses, but most of our communication with each other involves sound and vision. Not much communication between people involves the sense of smell. But now we are beginning to prove that we can influence each other with our smells — and, that we pick up these smells with a strange sex organ inside our noses!

When the swab hits your nose . . . that's amore!

a sense of smell

Scientists who study human anatomy have known for a long time about the 'olfactory epithelium'. 'Olfactory' means 'related to smell'. The olfactory epithelium is a patch of yellowish tissue high up in the 'ceiling' of the nose. Normally it is poorly ventilated, but when we sniff deeply, we pass lots of air over it. In this yellow patch, there are sensory cells specially adapted for smelling. Chemicals in the air enter the nose, excite the sensory cells, and then we get the sensation of 'smell'.

vomeronasal organ

There is also another area in the human nose that can be used to detect odours — but until recently, most scientists didn't believe it existed! This is the VNO, which stands for 'vomeronasal organ'. Fishes, birds and some mammals don't have a VNO, but it is very well developed in snakes and lizards.

It was first discovered by the Dutch anatomist, Ruysch, in 1703. It's about a centimetre inside the nose located right next to the wall that separates the nostrils, on the 'floor' of the nose. There's one in each nostril. It looks like a hollow tube, with only a very small opening (about one-tenth of a millimetre across) into the nose. Each VNO is very small and difficult to see. This might be why the vomeronasal organ fell out of favour, and soon anatomists didn't even believe it existed. By the 1930s, physiologists maintained that not only did we humans definitely not have a VNO, but there was no structure in the brain to process the information from any such organ. However, in 1991, a careful study revealed that 910 out of 1,000 people had an easily-found VNO. But the fact that we have one does not mean that it does anything.

Pheromones

Some scientists have long thought that the VNO might be involved in 'pheromones'. This was a word invented in 1959 by Doctors Karlson and Lusche. 'Phero' means 'to carry', and 'hormone' means 'to excite' — so 'pheromone' means hormones that travel through the air, from one creature to another. We humans are not usually consciously aware when we are 'smelling' pheromones.

Pheromones control mating preferences in hamsters, dominance relationships in male elephants, the timing of puberty, and they can even prevent implantation of fertilised eggs. Back in 1971, Martha McClintock showed that women living together tended to synchronise menstrual cycles — and that pheromones were almost certainly involved.

smells electrified

In 1994, Luis Monti-Bloch and his team from the University of Utah actually managed to thread very fine insulated electrical wires into the VNOs of volunteers. The researchers then wafted various smells up the noses of their volunteers, and looked for electrical activity in the cells of the VNO. These smells were various odourless chemicals from the skin of men and women. The volunteers had absolutely no conscious idea that they were responding to these smells — in other words, their olfactory epithelium, which smells perfumes and pollutants, did not trigger. The cells in the males' VNOs fired when they got female skin smells, and female VNOs responded to male skin smells. But the VNOs did not respond to skin smells from the same sex.

It was odd that the volunteers didn't consciously realise that their VNOs were being stimulated. We could waft a smell up their noses, and their VNOs could fire frantically with electrical activity — but all the volunteer experienced was a vague, generalised emotion of feeling fine.

smelly armpits

But in early 1998, an excellent experiment showed a more definite effect — that some female smells could trigger women's menstrual cycles. Kathleen Stern and Martha McClintock from the Psychology Department at the University of Chicago asked some 'donor' women to donate the smells in their armpits (via a pad that they wore for eight hours a day). They then exposed some 'recipient' women to these smells — which were completely odourless, as far the recipients' conscious brains were concerned. And of course, the two groups, donors and recipients, never met face to face.

The smells were taken from the donor women at two different times in their menstrual cycle. When the smells were taken before the donors ovulated, the recipients' menstrual cycles became shorter. But when the smells were taken right at the time of the donors' ovulation, the recipients' menstrual cycles became longer. The overall effect was to synchronise the cycle of the recipient with the cycle of the donor.

This was a pretty good experiment, but we're still not totally sure that the smells from one human can influence another human. For one thing, this effect happened in only 70% of the volunteers — so what's going on in the brains of the remaining 30%? In addition, our neuroanatomists have not yet proved that nerves from the human VNO travel to the relevant parts of the brain. With our current technology, the only way to do that is to get several corpses, add some dye to the VNO, wait a few months for the dye to migrate, and then very carefully cut open the brain and analyse it.

So if your boyfriend or girlfriend gets up your nose, it may not be their fault.

References

'Signals from pheromones go directly to the midbrain', *Australian Science Teachers Association*, vol. 47, no. 4, November 2001, p. 27.

Leinders-Zufall, Trese, *et al.*, 'Ultrasensitive pheromone detection by mammalian vomeronasal neurons', *Nature*, vol. 405, 15 June 2000, pp. 792–796.

Monti-Bloch, L., Jennings-White, C. & Dolberg, D.S., 'The Human Vomeronasal System', *Psychoneuroendocrinology*, vol. 19, nos. 5–7, 1994, pp. 673–686.

Monti-Bloch, Luis, *et al.*, 'The functionality of the Human Vomeronasal Organ (VNO): evidence for steroid receptors, *Journal of Steroid Biochemical Molecular Biology*, vol. 58, no. 3, 1996, pp. 259–265.

Motluk, Alison, 'Go for it, baby — being around women who breastfeed has an unexpected effect', *New Scientist*, no. 2340, 27 April 2002, p. 11.

Weller, Aron, 'Communication through body odour', *Nature*, vol. 392, 12 March 1998, pp. 126–127.

For well over a century, surgeons have been using anaesthetics, but exactly how they work still remains a mystery.

anaesthetic mystery

In the bad old days before anaesthetics were invented, surgical operations had to be done very quickly. The world record for a leg amputation is 15 seconds — by Dominique Larrey, who was Napoleon's chief surgeon. In fact, Charles Darwin was so disgusted by the suffering caused by surgery without anaesthetics, that he gave up a promising career in medicine — and that's how the Theory of Evolution came about.

We have been using anaesthetics for one and a half centuries, but there's still one great mystery — we have no idea how anaesthetics work.

'Anaesthesia' means 'no sensation' — which is really handy if your dentist is drilling in your mouth or your surgeon is inserting a metal pin in your leg. A general anaesthetic adds 'unconsciousness' to the 'lack of sensation'. Anaesthetists (or anesthesiologists as the Americans call them) know how many milligrams of what particular drug to give you per kilogram of your body weight — but they have no idea how general anaesthetics work at the basic human cell level.

we do know a few things

A general anaesthetic is carried in your blood to the nerves in your brain. The nerve cells stop *receiving* signals (so you don't feel any pain) and stop *sending* signals (so you don't thrash around in the middle of delicate surgery). We also know that a chemical needs to have certain properties to be a good general anaesthetic — it has to evaporate easily from a liquid into a gas, and it has to be soluble in fat. These properties cover real anaesthetics like halothane, or other chemicals like paint stripper and nail polish remover — which sometimes have nasty side effects, like death. Another thing we know is that the better a chemical dissolves in olive oil, the better a general anaesthetic it is — but again, we don't know why.

The restful slumber

anaesthetics go back a long way

The Ancients knew about the effects of alcohol, opium and hemp. Around 1800, the British chemist Sir Humphry Davy discovered that nitrous oxide could, after it made you silly, also make you unconscious. In 1842, the American surgeon, Crawford Long, used ethyl ether as a general anaesthetic. But Long didn't publish his results. Today, it is William Morton, an American dentist, who is given credit for the discovery. In 1846 he demonstrated that ether could be used as a general anaesthetic, by performing a tooth extraction in public. Within six months, the use of ether as an anaesthetic had spread around the world. Many people argued against anaesthetics, claiming that they interfered with God's 'natural order'. But the opposition faded away in 1853, when Queen Victoria used chloroform as an anaesthetic when she gave birth to her eighth child, Prince Leopold.

Discretion Advised

Even if you are fully and safely anaesthetised (and you don't wake up!) while you are being operated on, it seems that you can 'remember' what the operating theatre staff said during your operation.

You can't usually remember entire conversations, but you often do keep conscious or subconscious memories of things that happened, or were said, during the operation.

cell membrane sandwiches

While we still don't know *how* general anaesthetics work, we're pretty sure that it has something to do with the cell membrane of the nerve cell. Every cell in your body is wrapped inside its cell membrane — it acts as a barrier to keep the outside out, and the inside in. The cross-section of the cell membrane looks like a sandwich — with different layers of water-loving molecules and water-hating molecules. The pressure in between the different layers of the cell membrane is enormous — around 400 atmospheres, or roughly the pressure found 4 km below the surface of the Pacific Ocean.

We do know that the cell membrane has a whole bunch of channels which will let certain chemicals go in or out — so there are chloride channels, sodium channels, potassium channels and so on. Each channel passes right through the sandwich of the cell membrane, and the enormous pressure tries to close it up — but each channel is held open by cholesterol and other fats that are arranged into a fairly rigid liquid crystal. As various chemicals flow in or out of the channels of the cell membrane, they change the electrical charge on the cell membrane — and that process can switch the nerve cell on or off.

so how does anaesthetic work?

Here's where we run into the various theories about how general anaesthetics work.

One of the older theories states that anaesthetics can somehow slide into the sandwich of the cell membrane and interfere with the rigid liquid crystal that holds open the channel. If the shape of the channel changes, so too will the flow rate of various chemicals (e.g. sodium, potassium etc.) going in or out of the cell. This does explain why so many different volatile chemicals (from nail polish to petrol to halothane) will make you unconscious.

A more modern theory suggests that anaesthetics work by attaching themselves to little chemicals around the open mouth of the channel. These little chemicals can then open or close the channel, and again, this can lead to changes in the electrical charge on the cell membrane — and hopefully lead to anaesthesia.

Another theory says that anaesthetics alter the enormous 400 atmospheres of pressure inside the sandwich of the cell membrane, so that they can change the shape of the channels — and again, interfere with the electrical charge on the cell membrane.

We still don't really understand how anaesthetics work. But when we do, not only will we produce better anaesthetics which have fewer side effects, we could possibly even begin to understand what consciousness is — and until then, the experts will just have to sleep on it.

References

Anderson, Ian, 'Painkiller doubles as a brain saver', *New Scientist*, no. 2138, 13 June 1998, p. 24.

Holmes, Bob, 'Can you hear me?', *New Scientist*, no. 2132, 2 May 1998, p. 14.

Knight, Jonathan, 'Ten, nine, eight ... you're under', *New Scientist*, no. 2178, 20 March 1999, pp. 35–37.

The Indian Rope Trick is very famous. The earliest written account of it was given to us by the Arab traveller Ibn Battuta, in 1346, while he was travelling in Hangzhou in China.

indian rope trick

A performance would begin just before sunset, with the Indian fakir (a Hindu miracle worker) taking a length of hemp rope from a wicker basket and throwing it into the air. It would fly up in the growing darkness, until the top was no longer visible and then miraculously stay there. A slim young boy would climb the rope and be seen to vanish into thin air. He would refuse his master's call to come back down. The fakir would draw a lethal-looking knife, clench it between his teeth, and clamber up after the boy — and also vanish from sight. Then there would be a series of blood-curdling yells, and various dismembered limbs of the young boy would fall to the ground, followed by his head. The fakir would then shimmy down the rope, which would then collapse after him, throw a cloth over the scattered body parts, clap his hands, say *'Hey presto'*, and the young boy would leap up — miraculously reassembled.

how did the trick work?

The trick would be performed in imminent darkness, with hills or trees nearby. The magician would rely on a strong, thin black cord slung between two high points before the performance. Attached to the end of the hemp rope was a heavy black ball. When the rope was thrown in the air, the ball would hook over the black cord, so it could support the weight of the slim youth. Once the boy had reached the top he would secure the vertical rope to the horizontal black cord, which could then take the weight of the magician.

And the boy's dismembered body parts? They were bits of a shaved monkey dressed in clothes similar to the boy's, with a bit of red sauce splattered around for good measure. As conjuring tricks go, it's a classic.

The fakir could use a second method if there were no high points nearby between which to sling a rope. This method would involve a thin bamboo rod up the centre of the hemp rope.

Isamudin, an Indian magician, after several years of research, has recently revived the Indian Rope Trick. On 23 November 1997, he performed the classical version of the Indian Rope Trick before an estimated crowd of 25,000, in an open area on a beach far from the nearest tree.

getting the rope to stand

But if magic isn't your bag, David Acheson from Jesus College at Oxford, and Tom Mullin, a physicist at the University of Manchester, have worked out the physics of how to get your Indian Rope Trick off the ground.

Imagine that you have a stick resting vertically on the palm of your hand — and you don't want it to fall over. In 1908, Andrew

The old 'how to get a stick to stand upright' trick

Stephenson, a mathematician from the University of Manchester, proved that you could keep this stick upright simply by rapidly moving the palm of your hand up and down, instead of side to side. This is one trick that you can do at home, folks!

Now suppose you get two sticks and link them together with a little metal loop. Once again, you can keep the two sticks standing vertically by moving your palm up and down 'fast enough', while moving through a distance that is 'short enough'. Of course the tricky words are 'fast enough' and 'short enough'.

Luckily, in 1733, a Swiss mathematician called Daniel Bernoulli, worked out exactly what 'short enough' and 'fast enough' were. So David Acheson and Tom Mullin linked together three rods, each about 19 cm long. They used a DC motor to vibrate the bottom support up and down 40 times per second, through a total distance of around 2 cm. The set of three linked rods stood up perfectly straight. The scientists even pushed the three rods over 45° from the vertical — and they just wobbled right back up to the vertical again.

up and down is better

Why is it so, that this up-and-down motion works better than a side-to-side motion? When the motor is pulling the assembly downwards, it's actually pulling it down faster than it would fall by gravity alone. Think about one of the linked rods that would fall, thanks to gravity, off to the side. Pulling it down rapidly makes it straighten out and become vertical.

What does this have to do with the Indian Rope Trick? Well, a rope is made of many stiff physical fibres. Each individual fibre, if it is short enough and stiff enough, will stand up by itself. So you can look upon a rope as being a linkage of many short fibres. Unfortunately, the scientists couldn't get their motor to vibrate rapidly enough to keep a real rope standing vertically, let alone support the weight of a person.

Despite what scientists have discovered, the secret to the entire Indian Rope Trick remains the domain of Indian fakirs, magicians and conjurers.

References

Acheson, David & Mullin, Tom, 'Ropy magic', *New Scientist*, no. 2122, 21 February 1998, pp. 32–33.

Alonso-Sánchez, Francisco & Hochberg, David, 'Renormalization group analysis of a quivering string model of posture control', *Physical Review* (E), vol. 62, no. 5, November 2000, pp. 7008–7023.

Chowa, Carson C. & Collins, J.J., 'Pinned polymer model of posture', *Physical Review* (E), vol. 52, no. 1, July 1995, pp. 907–912.

Facts & Fallacies, 1989, Reader's Digest (Australia) Pty. Ltd. (Inc. in NSW), pp. 272–273.

How Is It Done?, 1990, Reader's Digest (Australia) Pty Limited. pp. 400–401.

Kalmus, Henry P., 'The inverted pendulum', *American Journal of Physics*, vol. 38, no. 7, July 1970, pp. 874–878.

Formula One racing is no longer just a physical skill — the engineering design that goes into these cars now makes this profession an 'art form'.

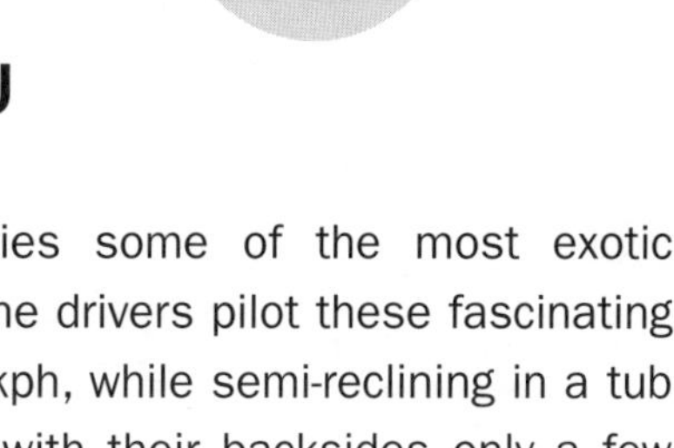

formula one

A Formula One racing car carries some of the most exotic engineering known to humanity. The drivers pilot these fascinating vehicles at speeds of up to 360 kph, while semi-reclining in a tub made of expensive carbon fibre, with their backsides only a few centimetres off the road. At full blast, a Formula One fuel pump delivers petrol faster than water flows out of your kitchen tap.

A Formula One car uses aerodynamics to generate, at full speed, a downforce of two and a half times its own weight, so that it will stick to the road really well. At 160 kph, they are generating their own weight in downforce — so they could theoretically drive upside down on the roof of a tunnel. The downforce exerted on the car means that it can corner at 5 Gs — but when the driver hits the bend, his head suddenly weighs 25 kg, and his 70 kg body now weighs a third of a tonne. The force of 5 Gs is enough to stop you from breathing. The drivers need supreme physical condition and concentration to ignore the G-forces and maintain their focus for the hour-or-so that it takes to cover 305 km.

where it all started

Following in the slipstream of the Formula One car is the domestic car. Many of its features (disc brakes, turbocharger, advanced tyre technology and sophisticated valve trains) are spin-offs from Formula One cars.

The very first car race took place between Paris and Rouen in 1894, the winning car averaging 16.4 kph. By 1971, the Italian Formula One Grand Prix was won at an average speed of 242 kph. The very first Formula One race was held in 1948. Today the Formula One Championship consists of a series of some 17 races, run every two weeks between March and October. After each race, the engineers have to tinker frantically with the car's design for the next non-negotiable race deadline in two weeks. And in each race, the car is substantially different from the way it was in the previous race.

Scientists working yet another angle for greater speed

Formula One today means that the engine has a capacity of 3 L or less, has 10 cylinders, and can't be supercharged. The car also has to weigh at least 600 kg, and have four wheels, only two of which are steered or driven.

These amazing engines generate some 600 kW at around 18,000 rpm. Compare this to your average Holden or Ford which generates about 140 kW at 4,000 rpm. Actually, by the 1990s, the engines were limited to only 12,000 rpm, because of friction in the valve train. But then Renault invented pneumatically-driven valves, which allowed the maximum engine speed to jump to 18,000 rpm.

'Psychopath' Behind the F1 Grand Prix Wheel

A 'psychopath' is 'a person whose behaviour is abnormal and may be antisocial'. Formula One Grand Prix drivers almost fit that definition. About a quarter of these drivers die a violent death — on the race track, or in their high-speed car, high-speed boat or Lear jet.

Dr Keith Johnsgard, a psychologist at San Jose State University, has studied many athletes, including GP drivers. Over the years, he and his team analysed a very strange animal. They found that these drivers have '*an extraordinarily high intelligence, a need for achievement unmatched among other athletes, very high assertive and dominance needs, and a high degree of tough-mindedness*'. They don't care what the current authority thinks, and are stubborn, self-assured and independent.

They have a great deal of self-control — they are not impulsive. Nor are they victims of some sort of death wish, because they direct their hostility outwards, not inwards to themselves. They are cheerful, frank, and happy-go-lucky. They never blame themselves, they don't give in to avoid trouble, and they don't get depressed — they sound just like Sylvester Stallone or Arnold Schwarzenegger.

The GP drivers are extraordinary for their '*unusually low needs for intimate interpersonal relationships, their non-conformity, their fierce independence and their remarkable freedom from guilt*'. GP drivers are exhibitionists, and need attention, especially from women, though they are shallow in their dealings with women. They have no conscience, and they don't form close relationships with people.

They don't want close friends, and they don't need someone to talk with and to understand their feelings. They don't ask for advice, and they don't give it. They seem to follow the motto of that hard man, Henry Ford: '*Don't explain, don't complain.*'

Psychologists claim that these drivers showed a '*lack of serious psychopathology, but ... mildly psychopathic personalities*'.

GP drivers are also '... *radical, liberal, critical and free-thinking*'. So they're not all bad.

But in the game they play, the odds of dying a violent death are one in four. Who can blame them for a slightly different personality, if it keeps them alive? Anyway, you can bet GP drivers don't give a 'flying spin' for what the psychologists think!

what makes them tick?

Each car has about 9,000 different components. The body and the chassis are made from carbon fibre, which is four times stiffer and five times stronger than steel. The carbon fibre steering wheel alone costs $120,000.

Each car has about one and a half kilometres of wire, integrating the data from some 120 sensors that glean information such as the angle of the rear wing, the brake temperature, the oil pressure and the tyre pressure. These vital statistics are constantly relayed back to the crew in the pits. The software, which integrates the data from the sensors and manages the engine and the gearbox, comprises some half a million lines of code, which took some 20 person-years to write. The gearbox can have up to seven different speeds — and if the timing of the changing of the gears is off by even a few thousandths of a second, the gearbox will self-destruct.

money to burn

It takes a lot of money and brain power to roll one of these Formula One babes out onto the track.

McLaren, for example, has a budget of about $500 million per year and employs 350 people. They have even constructed 3D digital replicas of each racetrack, in order to test the engine before each race.

McLaren invented a new braking system (progressive electro-hydraulic power brakes) which dramatically shortened the braking distance — but it was banned. They also invented a unique rear differential, to better feed the awesome power to the back wheels without spinning — but it too was banned. In fact, one engineer estimated that half of his 15-year career had been spent on developing engineering wizardry that was now illegal.

The job of the scientist is to *discover* phenomena that are already there, but currently unknown to humans. The job of the engineer is to *create* — to design and build something that has never been built before. You could say that today, in the design of the Formula One car, engineering comes close to art.

References

Benson, Andrew, 'Primal', *Wheels Magazine*, April 2001, pp. 112–115.

Hoyle, Simon, 'Formula One — powered by IT grunt', *The Australian Financial Review (Information)*, 1 March, 2002, p. 41.

Tilan, Andrew, 'Formula 2001', *Wired Magazine*, March 2001, pp. 130–143.

How can geckos stick to all kinds of surfaces? Do they have magic suction caps at the ends of their feet, or is it something else?

atomic gecko

If you have ever been to the tropics, you will have heard geckos clicking noisily, and you will have seen them scuttling up walls, and even hanging out on ceilings. Some encyclopaedias state that they stick to the ceiling with the help of tiny suction caps on the ends of their feet. Other books claim that the tiny hairs on the ends of geckos' feet adhere to microscopic rough spots on the surface they are climbing. But don't believe everything you read.

Geckos need something really unusual to explain their incredible 'sticking' ability. They can splash through a shallow puddle of water, then run through very fine dust, and then skip across an oily floor before finally running up a vertical wall and hanging off a ceiling. No commercially available sticky tape will stick after this kind of treatment.

In fact, geckos stick to the ceiling using atomic forces.

smooth gecko

Geckos are a type of lizard, and there are about 750 different species of them. They tend to be small (3–15 cm long), are active at night, and have a short stout body with a large head. They can survive anywhere from suburbia to the jungle or the desert. Geckos have the most highly developed hearing of any of the lizards. And they are the only lizards that make any kind of noise other than a 'hiss'.

For a century, people have been trying to work out how geckos really do stick to something as smooth as glass. But it took a huge team effort led by Robert Full from the Department of Biology at the University of California at Berkeley to figure it out. At Berkeley he had help from colleagues in the Department of Electrical Engineering and Computer Science and the Department of Integrative Biology. He also had outside help from the Department of Mechanical Engineering at Stanford University, and from the Department of Biology at Lewis and Clark University in Oregon. It took many different disciplines to understand something as simple as how a gecko sticks to a smooth surface.

Gecko Tails

There are about 100 species of gecko in Australia. Some of them do weird stuff with their tails.

The Fat-Tailed Gecko uses its short, flattened tail to block the entrance to its burrow in the ground. This protects it from attackers, and also conserves moisture.

The Chameleon Gecko can drop its tail (which is full of fat, and so makes an attractive morsel) when it's attacked. The tail makes a loud and distinctive squeaking noise, to attract the predator.

The Northern Spiny-Tailed Gecko has glands inside its tail. They can squirt a sticky, irritating liquid some 30 cm at an attacker.

The Giant Cave Gecko has a long tail that seems to be able to stick to the ceiling. I wonder how it works?

how does a gecko stick?

Suction has nothing to do with the gecko's legendary stickiness, because its feet will stick in a vacuum. Friction is not the answer either, because they stick to low-friction surfaces such as silicon. 'Microinterlocking' with tiny rough spots is not the answer, because they stick to super-polished, super-smooth glass. And electrostatic attraction is not the answer either, because the feet stick when the air is full of charged particles.

A gecko has four feet. Each foot has about half a million tiny hairs on the end. These hairs are very small. A human hair is about 70 microns across (a micron is one millionth of a metre). The hairs on the end of a gecko's foot are about 7 microns across and about 30–130 microns long.

But if you think that these hairs are really small, have a look at them through an electron microscope, and you'll see that things get even smaller. Each of these half a million microscopic hairs has several hundred smaller hairs coming from it — each one being about 0.2–0.5 microns across. These itty-bitty hairs, called spatulae, are roughly the same size as a wavelength of light. They are extremely small indeed.

The research team was able to build, and then attach, a microscopic force-measuring device to one of the half a million

The ultimate in climbing apparel

hairs — and then measure the force needed to pull it off a surface.

First, they discovered why the toes on a gecko have a peculiar way of curling as they stick and then uncurling as they peel their feet from a surface. It's a bit like removing sticky tape. The hairs can generate a lot of grip. They are lined up in such a way so they need only a small amount of force to pull them off — so the gecko won't get exhausted as it peels its feet off the ceiling.

Second, the force measurements taken from the single hair on the gecko's toe, suggest that geckos stick with atomic forces! These are called 'van der Waals Forces', after the 19th century Dutch physicist who first described them.

how do these forces work?

Imagine that you have a nice round atom, with a central core or nucleus, surrounded by a neat symmetrical cloud of electrons. Over a long period of time, say a second or so, the cloud of electrons is perfectly symmetrical. But over a much shorter time, say a trillionth of a second, the cloud of electrons might be lopsided, with more of the electrons (say) on the left side of the atom, making the left side of the atom temporarily more negative. This induces a tiny positive charge in another atom nearby. The opposite charges attract — and there's your atomic attraction. Of course, in real life, van der Waals Forces are more complicated, but this is basically how they work.

We now believe that the really tiny hairs on a gecko's foot are so small (sometimes smaller than a wavelength of light) that they can get close enough to the surface to stick by direct atomic or van der Waals forces.

What use is all this high-powered science? After all, our current technology can't create a packed array of a billion microscopic spatulae. But, now that we know how geckos do it, we might be able to invent the first dry adhesive. And because geckos have sticky feet that work perfectly well after they run across dirt we might be able to invent large self-cleaning adhesive surfaces, that have a tremendous grip, but are still easy to remove. One day, climbers could use gecko-technology gloves and shoes to stick to rock faces — and Mr Mission Impossible, Tom Cruise, wouldn't slip off vertical stone walls anymore.

Spiderman Comes

In mid-2003, Andre Geim and his colleagues from the University of Manchester made the first Gecko Tape. The tiny hairs that do the 'atomic gripping' are not made of keratin (like in a real gecko), but of a flexible, wear-resistant plastic called polyimide. Like keratin, it repels water. One square centimetre of Geim's tape will hold 3 kg, so 30 cm^2 could easily hold 90 kg. If you had a glove made of this Spiderman Tape, you could hang yourself from a ceiling or on a wall quite a few times before the tape loses it 'stick'.

It's not nearly as good as a gecko, but it's a start. Real geckos can run across finely powdered dirt, and then up a wall. No glue on the market today can do that.

In the future, more durable versions of this super-strong Pseudo-Gecko Tape could help rock climbers grip onto rock faces, move delicate or huge parts in a vacuum, stick wounds together, or give rubber car tyres a truly incredible grip.

References

'Gecko toes tap intermolecular bonds', *Science News*, vol. 158, July 15 2000, p. 47.

Autumn, Kellar, *et al.*, 'Evidence for van der Waals adhesion in gecko setae', *Proceedings of the National Academy of Sciences*, vol. 99, no. 19, 17 September 2002, pp. 12252–12256.

Autumn, Kellar, 'Adhesive force of a single gecko foot-hair', *Nature*, vol. 405, 8 June 2000, pp. 681–682.

Geim, A.K., *et al.*, 'Microfabricated adhesive mimicking gecko foot-hair', *Nature Materials*, no. 2, July 2003, pp. 461–463.

Svenson, Nic, 'Gecko's sticky secret is a hairy one', *The Sun-Herald*, 15 June 2003, p. 36.

Wilson, Steve, 'Geckos, the eyes have it', *Australian Geographic*, July 2000, pp. 72–89.

Marriage is full of mysteries. Could a mathematician come to the rescue by helping us work out the best age at which to marry?

the maths of marriage

The institution of marriage goes way back. Over 5,000 years ago, the Sumerians had laws which regulated marriage. But these laws were still 'rough-house' rules. The 'Best Man' was so called because he would help kidnap the bride, and then fight off any relatives who tried to rescue her. The Romans thought this a bit uncivilised. They went one step further, enacting laws that made sure that both the groom and the bride entered into the marriage of their own free choice.

The very earliest marriage certificate on record was found in a bundle of Aramaic papyri, and is some 2,500 years old. It was discovered in the ruins of a Jewish garrison, that had been stationed at Elephantine in Egypt. It's more of a 'contract' than a 'marriage certificate', as it documents that the groom landed himself a healthy 14-year-old girl bride in exchange for six cows.

Many people would think that 14 years of age is a little too early to get married. If you make *'the decision'* too early, you would be out of the running, and not be available to marry the person who may be best suited to you. But if you wait too long before making *'the decision'*, you might have let that special person slip through your fingers into someone else's hands.

when to marry

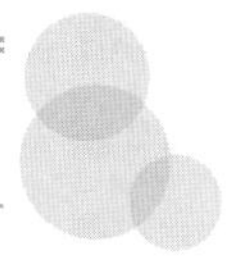

There are many ways to look at this problem. For example, let's look at it with regard to how attractive the man is. On one hand, as a man gets older he becomes less virile, but on the other hand he becomes more mature. Looking at it this way, the best time to get married is when the decreasing virility is balanced by the increasing maturity. And there are other points of view. You could look at the when-to-get-married problem from the viewpoint of the Law of Diminishing Returns or the Law of Supply and Demand.

But the mathematician, Dr D.V. Lindley, discussed it from an entirely different point of view in his book called *Making Decisions*. You see, each time you (a single person) meet a potential marriage partner, you make what is usually an irreversible decision — you either marry the person, or you look for somebody better. How do you know when you've found the person who is best for you?

Mathematicians have a game called *Googol* that may solve this 'marriage problem'. A friend writes down 100 different numbers on 100 different pieces of paper. They show you the numbers, one at a time. You have to guess when you have been shown the biggest number out of all the numbers that your friend recorded.

Lindley worked out that the best time to make your decision is roughly one-third of the way into your period of availability.

(Actually, if you want to be exact, you divide your total time by a number that mathematicians call the exponential *e*, which is roughly equal to 2.718). With 100 numbers in *Googol*, you would make your guess after about 30 numbers, once you've seen the biggest number so far.

For women, the period of marriage availability is about 30 years, from 16 to 46 years of age. One-third is roughly 11. Add 11 to 16 and you find that the 'mathematically correct' age for a woman to get married is 27. You might say that men have a longer period of marriage availability, from 16 to 60 years of age, which is a 44-year window. One-third is roughly 16. Add that to their 16-year starting time, and you find that the mathematician's time for a man to get married is 32 years of age.

googol not good

These numbers are interesting, but they're quite different from the average ages at which men and women do get married. So either our rather simplistic mathematical model is hopelessly wrong, or

people are not logical — or both. It's easy to see that our simplified model of marriage ignores many factors: not all married couples want babies; easy access to contraception means that people can have 'baby-free' sex outside marriage; increasing levels of education tend to delay marriage, and sometimes the choice is not irreversible — people do go back to marry people from the past.

And if you want to be really logical, women should marry men who are about six years younger than them. That way, not only are the different sexual appetites more closely matched, but the two marriage partners will probably die about the same time.

But no matter how complicated marriage is, it will probably always be with us. After all, as George Bernard Shaw said: *'Marriage is popular because it combines the maximum of temptation with the maximum of opportunity.'*

As far as the maths of marriage are concerned, sometimes the numbers don't add up. But then again, sometimes the result is greater than the sum of its parts.

References

Ferguson, Thomas S., 'Who solved the secretary problem?', *Statistical Science*, vol. 4, no. 3, 1989, pp. 282–296.

Lindley, D.V., *Making Decisions*, (paper), 1985, pp. 144–149.

Matthews, Robert, 'Decisions, decisions', *New Scientist*, no. 2113/2114, 20/27 December 1997, pp. 58–61.

Quine, M.P. & Law, J.S., 'Exact results for a secretary problem', *Journal of Applied Probability*, vol. 33, 1996, pp. 630–639.

Swets, John A., Dawes, Robyn M. & Monahan, John, 'Better decisions through science', *Scientific American*, October 2000, pp. 70–75.

Another way of restarting the heart might come from 'down under' — by injecting fluid into the penis.

intravenous penis

The heart is not the organ of love — it's a pump. And what a mighty pump it is! Although it's only the size of your clenched fist and weighs only a few hundred grams, in an average lifetime it will pump around 180,000 tonnes of blood — roughly the weight of a few of those enormous nuclear-powered aircraft carriers that can rush around the world at 80 kph non-stop for five years.

But the heart has one weakness — it's not very good at restarting itself, if it happens to stop. All sorts of emergency resuscitation methods have been invented to keep the heart beating and prevent it from stopping. And the newest one might give a bit of a shock to the blokes — because it involves injecting fluids directly into the penis.

the heart of the matter

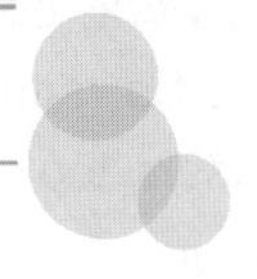

The average person has about five litres of blood. It circulates through some 100,000 km of blood vessels. The heart pumps about five litres of blood every minute.

There are various situations in which you can lose a lot of blood from your circulatory system. These include burns, the loss of a limb, internal bleeding into your abdominal cavity, obstruction of your intestines, and virtually any injury that involves the tearing open of a blood vessel.

For many years, the cornerstone of resuscitation has been the need to gain access to some kind of blood vessel (usually a vein), and then to shove in fluids and drugs quickly. The fluids, which could be anything from blood to a saline solution or coconut juice, would give the heart something to pump, and the drugs would help the heart pump more efficiently.

But the terrible irony was that the very time you needed to have access to a blood vessel, was when the blood vessel had collapsed to its smallest size. You see, when a person has lost a lot of blood, the veins are pretty flat and it's almost impossible to find them. There are emergency manoeuvres that a medico could use such as stabbing a needle directly into the bone (which has a rich blood supply) or even doing a 'cut down'. This involves cutting the skin with a knife, and working your way down until you find a vein, and then using the vein as your access into the circulatory system — but this procedure takes a lot of time.

resuscitation via the penis

Recently, a team of surgeons from the University of Queensland came up with a new technique — resuscitation via the penis. This can work, because the penis has a special type of blood vessel called a 'sinusoid'. Practically everywhere else in the body, blood vessels look basically like pipes — either straight or branching. But in the penis (and in the clitoris and in the nose) there is sinusoidal tissue. In sinusoidal tissue, the blood vessels are arranged like a sponge and go every which way. In the case of the penis, this is the hydraulic mechanism that allows for erection. The blood comes in at a certain rate, engorging the sinusoidal tissue, and goes out at a slightly slower rate, thus leading to erection. It turns out that the sinusoidal tissue is made up of a stiff fibrous tissue that holds the blood vessels open — even when the circulating blood volume is low. So even when the patient is in severe circulatory collapse, and the veins have all gone flat, you can still get access to the blood vessels via the penis — or so the surgeons thought.

In a series of studies, the team dumped varying amounts of liquids into the penises of dogs. They first removed a lot of the circulating blood from the dogs, and then stabbed the penis with a

19-gauge needle (only a little smaller than a match). They then poured in vasoactive drugs and fluids through the needle. The experiment worked. These drugs appeared in the body's general circulation within one minute. They dumped fluids in at rates up to 67 ml per minute. This was enough to bring the dogs back to a normal circulatory state.

So if future tests prove successful, this direct injection emergency procedure into the penis could one day be used on men. And there is one consoling factor for the blokes who might need this new and unusual form of resuscitation, they won't have to ponder whether they'll give consent, because they'll probably be unconscious.

References

Guyton, Arthur C. & Hall, John E., *Textbook of Medical Physiology*, Chapter 24: 'Circulatory shock and physiology of its treatment', W.B. Saunders Company, 1996, pp. 285–294.

Hurren, J.S. & Dunn, K.W. 'Intraosseous infusion for burns resuscitation', *Burns*, vol. 21, no. 4, 1995, pp. 285–287.

Nicol, D., Watt, A., Wood, G., Wall, D. & Miller, B., 'Corpus cavernosum as an alternative means of intravenous access in the emergency setting', *Australian and New Zealand Journal of Surgery*, vol. 70, 2000, pp. 511–514.

The wedding-ring finger isn't just reserved for that special band of gold, it can also help you find the real testosterone tigers!

sexy finger

The wedding ring has a long tradition. Nearly 5,000 years ago, in 2800 BC, the Ancient Egyptians from the Third Dynasty of the Old Kingdom used a ring in their wedding ceremonies. Since then, wedding rings have been used by many different societies. A wedding ring usually goes on the fourth finger, the one next to the little finger. Well, it might be a coincidence, but the length of your fourth finger can tell you heaps about the levels of your sex hormones.

So hold up your right hand and have a look at your second or index finger — the one that you point with. Then compare it to your fourth finger — the one next to the little finger. Here's the big question — which finger is longer, the fourth finger or the second finger?

Licence to thrill!

Or
Is that a big 4th finger you have . . . or are you just happy to see me?

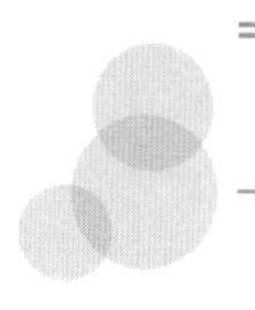

the long and the short of it

Here's the deal for men. If the fourth finger of your right hand is *longer* than your second finger, then you are probably a super spunk, a stud-muffin, a testosterone tiger. On average, you (male with the longer fourth finger) will have a higher sperm count, a higher level of testosterone, and a lower incidence of germ cell failure, which manifests itself in a zero or a very low sperm count. You'll probably be good fatherhood material. But for women, it's the other way around. Women whose fourth finger is *shorter* than their second finger will have higher levels of various female hormones such as oestrogen, prolactin and luteinising hormone.

This study was carried out by a team headed by Dr Manning from the Population Biology Research Group at the School of Biological Sciences at the University of Liverpool. The team was looking for a relationship between the length of the fourth finger and the various male and female sex hormones. They looked at some 800 people in all — 400 males and 400 females in the local Merseyside area — ranging from toddlers of two years of age to adults of 25 years of age. On average, the men were likely to have longer fourth fingers, and the women, longer second fingers.

Why would anybody look for a link between the levels of your sex hormones and the length of your fingers? Well, the link lies in your DNA — specifically how your DNA controls where your various body parts will go.

In 1984, three biologists working with fruit flies found some genes in DNA that they called the 'homeobox'. One of them, Edward B. Lewis, had been working with fruit flies since the 1940s. The other two, Christiane Nüsslein-Volhard and Eric F. Wieschaus had begun their work in the 1970s. The three biologists discovered that the genes in the homeobox of fruit fly DNA control the extraordinary change from a single fertilised egg into a fully-fledged fruit fly.

Sport and Fingers

If you are a top-level male soccer player, you are more likely to have a longer ring finger than a lower-level male soccer player.

Dr John Manning came to this conclusion after he and his team checked out the hands of 304 male professional soccer-players and their coaches — in upper and lower divisions. This 'longer fourth finger' also applied to winners on the squash court, running track, rugby field, judo mat, hockey pitch and in the Olympic swimming pool. They all had better coordination, physical fitness and spatial judgment.

Dr Manning said: *'Our fingers tell us how much testosterone we have been exposed to before birth.'*

the homeobox gene

A fruit fly has a head at one end, a central body, and a bunch of legs and wings. In 1995, the three biologists were awarded the Nobel Prize for Physiology or Medicine for their research into the homeobox. Homeobox genes are arranged in a sequence that is almost an exact replica of how the body parts in the finished fly are arranged. For example, the DNA at one end of this gene controls the making of the fly's head and upper thorax. And the DNA at the other end control the making of the fly's lower thorax and abdomen.

We are still learning about the homeobox gene — but we do know that the homeobox is found in the DNA of every creature tested so far. Each of these living creatures has a head-to-tail organisation — and that goes for humans as well. The bit of the homeobox that controls the manufacture of your hands also controls the manufacture of your genitals. More specifically, the ends of your arms (the hands and fingers) and the end of your trunk (the genitals) are controlled by the same bit of DNA.

In other words, the homeobox governs your reproductive potential! We know (from this study) that this arrangement is definitely locked in by the age of two. And we are pretty sure (from our study of embryology) that it may even be locked in a few months after conception. The University of Liverpool scientists were using this link in their research. And they were right. The longer fourth finger makes a man a super-stud, while the longer second finger makes a woman super-foxy. So it's not just destiny that controls our ends, our DNA has a big hand in it as well.

growing new body parts

The homeobox makes sure that your arms go at the top of your chest (not the bottom), and your hands go on the end of your arms (not your legs). When the homeobox goes wrong, it can cause diseases. But if we learn how to tweak the homeobox, we could stop these problems, and, perhaps grow new body parts.

Today, St Cosmos and St Damian are the patron saints of surgeons. They got this exalted status from a famous painting by the 15th-century Spaniard, Alonso de Sedano. He painted what looks like the first surgical transplant operation. It depicts Drs Cosmos and Damian performing a miracle — they are removing a white cancerous leg, and replacing it with a black leg from a dead Moor (a Moslem of African descent). That was a pretty impressive trick — but wouldn't it be a lot easier and safer if instead, you could just grow a new leg, where the old leg used to be?

One day we might be able to do just that, thanks to the three biologists whose research into the fruit fly discovered the homeobox. The fruit fly body is made up of repeating segments that are customised to perform special tasks. There are three segments that each have a pair of legs, one segment that sprouts wings, and another segment for the antennae on the head. But every now and then, biologists have seen sporadic errors in how the segments got made and arranged. They've seen the occasional mutated fruit fly with legs growing on its head where the antennae should be. The cause was found in the homeobox.

We think that the homeobox is also involved in a few minor human malformations. For example, most of us have 12 ribs on one side and 12 ribs on the other. But some of us have 13 ribs on one side and 12 on the other — and that's probably a homeobox 'error'.

Another human malformation is craniosynostosis. When a baby is born, the flat plates of bone that make up the skull actually overlap each other to make the head smaller, so that the baby can squeeze through the birth canal. These bones gradually grow closer to each other during the early years of life, and fuse together. With craniosynostosis these gaps between the flat bones fuse shut too early. The baby's brain needs to keep growing, but the skull won't expand — and things can get really messy. To treat this condition, the bits of bone that are joined together are removed, and a polyethylene film barrier is inserted to stop the bones from fusing back together again. We think that craniosynostosis, this premature fusion of the gaps between the bones of the skull, is related to an error of the homeobox.

We also think that retinoic acid, a chemical related to vitamin A, interferes with the homeobox. That's why pregnant women should not take too much vitamin A.

Now that scientists have mapped all of the human DNA, we know a lot more about the homeobox region. Perhaps one day we'll be able to re-grow limbs that were lost in accidents, instead of fitting artificial limbs.

But people being what they are, there'll be plenty of interest in seeing if we can extend the length of the fourth finger.

Legs for days!

Way too many legs . . .

Foot and Penis Size

We know that there is a link between a man's testosterone levels and the length of the fourth finger on his right hand. But what about the enduring folk myth that there is a link between foot size and penis size? The first thing to realise is that like everything in the body, many different factors control penis size.

Heinrich Loeb was one of the first to study this field of foot/penis size. He published his results in 1899. He measured an incredible array of different body dimensions besides penile length, such as body height, penile circumference and even the volume of the urethra. His results showed that many different factors affect penile length — and that body height accounts for 20% of this length. He did not specify what accounted for the other 80%.

In 1993, Dr Kerry Siminoski and Dr Jerald Bain published their follow-up results in the Annals of Sex Research. They measured height and 'stretched penile length' of 63 'normally virilised' men. (Note that 63 is definitely not a huge sample size.) They also recorded shoe size, which they later converted to foot length. They found that overall height and foot size each separately accounted for about 7% of penile length, leaving about 86% not accounted for.

The reason why their results are different from Loeb's 1899 results are probably because each study looked only at a small number of men. Also, neither study looked at probably the most important factor — the influence of the male sex hormone, testosterone. Dr Manning's study in 1998, however, did look at testosterone and the length of various fingers — but it did not look at 'foot size'.

So the straight answer in the early 21st century to the question of 'penis size' and 'foot size' is simple — we still don't know. But we definitely do know one thing about men with large feet — they wear large shoes.

References

Adler, Bob, 'In the mood', *New Scientist*, no. 2214, 27 November 1999, p. 18.

De Robertis, Eddy M., Oliver, Guillermo & Wright, Christopher V. E., 'Homeobox genes and the vertebrate body plan', *Scientific American*, July 1990, vol. 263, no. 1, pp. 26–32.

Kondo, Takashi, Zákány, József, Innis, Jeffrey W. & Duboule, Denis, 'Of fingers, toes and penises', *Nature*, vol. 390, 6 November 1997, p. 29.

Laughon, Allen S. & Carroll, Sean B., 'Inside the homeobox', *The Sciences*, March/April 1988, pp. 42–49.

Manning, J.T. & Bundred, P.E., 'The ratio of 2nd to 4th digit length: A new predictor of disease predisposition', *Medical Hypotheses*, vol. 54, no. 5, 2000, pp. 855–857.

Manning, J.T., Scutt, D., Wilson, J. & Lewis-Jones, D.I., 'The ratio of 2nd to 4th digit length: a predictor of sperm numbers and concentrations of testosterone, luteinizing hormone and oestrogen', *Human Reproduction*, vol. 13, no. 11, 1998, pp. 3000–3004.

Mould, R.F., *Mould's Medical Anecdotes*, Adam Hilger Ltd, 1984, p. 96.

Williams, Terrance J., *et al.*, 'Finger-length ratios and sexual orientation', *Nature*, vol. 404, 30 March 2000, pp. 455–456.

There's an infinity of numbers, but the universe seems to be in love with the number '1'.

benford's law

In our modern world we are surrounded by numbers. These numbers include speed limits, prices of goods, ages of friends and distances to travel. And if you examine these numbers, you'll discover a very odd phenomenon — nature has a soft spot for the number 'one'.

This was first discovered around 1881, when the American astronomer Simon Newcomb wrote a letter to the *American Journal of Mathematics*. He had noticed something odd about books of logarithms. Logarithms were used by scientists and engineers to multiply big numbers, before cheap electronic calculators appeared in the 1970s. If you wanted to multiply two numbers together, you'd look up their logarithms in a book, add them together, look up the anti-logarithm of that number, and there was your answer.

Newcomb noticed something odd about these books of logarithms — the early pages were much dirtier than the back pages. This meant that scientists and engineers spent a lot of time dealing with numbers beginning with 1, less time with numbers beginning with 2 and so on. In fact, Newcomb proposed a law which stated that the probability that a number will begin with the digit N is equal to log(N + 1) — log(N). But the mathematicians weren't interested.

benford's bent

The first real burst of interest was generated in 1939 by Frank Benford, a physicist working with the General Electric Company in the USA. He accidentally came across the effect that Newcomb had mentioned. But Benford probed a little bit deeper. He looked at a huge sample size, much bigger than Newcomb's. He analysed over 20,000 numbers retrieved from collections as obscure as the drainage areas of rivers, to stock market figures and the various properties of different chemicals. Again he showed that about 30% of numbers began with the digit '1', 18% with '2', all the way down to 4.6% of numbers starting with the number '9'. The law is now called Benford's Law in his honour, even though he wasn't the first to discover it. But Benford's Law doesn't apply everywhere.

First, you need to have a big enough sample size so that patterns can show themselves. For example, you almost certainly won't find Benford's Law in your last two telephone bills. Second, you don't want numbers that are truly random. By definition, in a random number, every digit from 0 to 9 has an equal chance of appearing in any position in that number. And third, you don't want numbers that are the complete opposite of random, and are tightly controlled. So if you deal with numbers that have artificial limits, such as the prices of petrol in a capital city, you won't find Benford's Law. In this situation market forces lock the prices of petrol into a narrow range.

Why it Works

Nobody understood why Benford's Law actually worked until 1996, when Theodore Hill, a mathematician from the Georgia Institute of Technology in Atlanta, began to unlock the mystery.

Everything we measure in the Universe is caused by something else, and it follows some kind of Law of Mathematical Distribution. For example, the height of trees in a forest follow the classic bell-shaped Gaussian curve. The varying distance between the Earth and the Moon follows a kind of regular up-and-down rhythm. So each of these measurements are governed by different Laws of Mathematical Distribution.

But Theodore Hill showed that if you take data from different Laws of Distribution and combine them, you end up with a strange sort of Law called the Distribution of Distributions — and this turned out to be Benford's Law.

somewhere in between

But Benford's Law does apply to numbers somewhere between totally random and totally constrained — such as the amounts on monthly electricity bills in the Solomon Islands.

The numbers that appear in accountancy tables and in the balance sheets of companies follow Benford's Law. This was

discovered by Mark Nigrini, in his PhD thesis written in 1992. He showed that sales figures, buying and selling prices, insurance claim costs and expenses claims should all follow Benford's Law.

Dr Nigrini gave his accountancy students an assignment. They were supposed to examine some real sales figures from a real company, and see how many of them started with the digit '1'. One of his students used the figures from his brother-in-law's hardware shop. The sales figures were astonishing — 93% began with the digit '1', none of them began with the digits between '2' and '7', four began with the digit '8', and 21 with the digit '9'. According to Benford's Law, the brother-in-law was a crook who was cooking the books — and Benford's Law was right.

number fraud

Nigrini calls the use of Benford's Law to find fraud 'digital analysis'. According to Nigrini, *'It's used by listed companies, large private companies, professional firms and government agencies in the USA and Europe — and by one of the world's biggest audit firms.'*

People who are not accountants are also interested in using Benford's Law to uncover other types of trickery. Mark Buyse and his colleagues from the International Institute for Drug Development in Brussels think that Benford's Law could expose data that had been faked in drug trials. Peter Schatte, a mathematician from the Bergakademie Technical University in Freiberg is using Benford's Law to organise space efficiently on computer hard drives.

Who would have thought that Simon Newcomb would start the ball rolling in 1881, by having a close look at some dirty books?

References

Browne, Malcom W., 'Following Benford's Law, or looking out for no. 1', *The New York Times*, 4 August, 1998.

Matthews, Robert ,'The power of one', *New Scientist*, no. 2194, 10 July 1999, p. 27.

Stewart, Ian, 'The half-life of a dirty book', *New Scientist*, no. 1872, 8 May 1993, p. 12.

Most kids have some kind of hobby — a sport, stamp collecting or computer games. But David Hahn, who lived in Michigan, USA, had a scientific hobby — chemistry. And so he tried to build a nuclear reactor.

radioactive boy scout

At the age of ten David was given a book called *The Golden Book of Chemistry Experiments*. Something clicked, and by the age of 12 he had mastered his father's university-level chemistry books, and by 14 years of age, he had made nitroglycerine. David's father thought that his son needed a stabilising influence, so he advised David, who was a boy scout, to aim for the level of Eagle Scout — which needs a total of 21 merit badges. Some merit badges (like citizenship studies, first aid and personal management) are compulsory, while some (from the study of American business to Zoology) are chosen by the scout. David chose the Merit Badge in Atomic Energy studies.

To get this badge, you have to know about nuclear fission, know who the important people in the history of atomic energy are, and make a few models of some atoms, and other stuff.

Radioactive Boy!

David built a model of a nuclear reactor with some tin cans, drinking straws and rubber bands, and earned his merit badge in Atomic Energy on 10 May 1991 (when he was 14 years and 7 months old). Then he decided to aim higher.

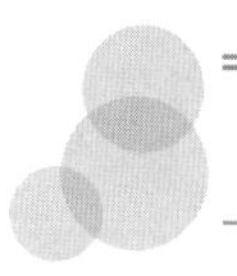

the neutron gun

Atoms have a core of positively-charged protons and neutral neutrons. Some of the bigger atoms (like uranium, for example) have unstable cores. If a neutron hits this core, it splits into two smaller atoms and a neutron or two — and also gives off a huge amount of energy.

So David decided to make a neutron gun. He pretended to be a Physics lecturer, and got lots of professional help from industrial companies, the American Nuclear Society and the Nuclear Regulatory Commission. He found out that he could get radioactive americium-241 from household smoke detectors — so he bought 100 broken ones at a dollar each. The friendly customer-service representative told him exactly where the americium-241 was, and how to remove it from its inert gold matrix. He then put his tiny pile of americium-241 inside a hollow lead block, and drilled a small hole in it. As americium decays, it gives off α-particles. When α-particles hit aluminium foil, the aluminium gives off neutrons. So he placed a strip of aluminium in front of the hole in the lead block where the α-particles came out, and bingo — he had a neutron gun.

He had discovered that the cloth mantles in gas lanterns are covered with thorium-232 (because thorium is very resistant to high temperatures). He also knew that if you hit thorium-232 with enough neutrons, the thorium-232 would turn into uranium-233. So he bought thousands of gas mantles, and used a very hot gas flame to turn them into thorium ash. How did he purify the thorium? Simple — he bought a few thousand dollars worth of lithium batteries, cut them open, and did some simple chemistry to concentrate the thorium. But alas, the effort was wasted. His neutron gun didn't have enough grunt to turn thorium-232 into lots of uranium-233.

Nuclear Reactor

I once thought that nuclear reactors generated energy by somehow turning one of the nuclear forces directly into electricity — and that no moving parts were involved.

But no, today's nuclear reactors are incredibly primitive. If you get a few lumps of Uranium-235 very close to each other, they will go Bang! If you pull them back a little way from each other, they will just get very hot (and spray out heaps of radiation as well). If you run liquid water over the Uranium-235, the heat turns it into steam. You can use this steam to spin a turbine, and make electricity.

Today's nuclear reactors are just steam engines powered by heat. The heat comes from Uranium-235, but it could just as easily come from burning coal, bank notes or paintings.

making radium

Time for plan B. Radium delivers heaps of α-particles, and he had been told that if you blast these α-particles onto beryllium, you get enormous numbers of neutrons. But how could he get some radium? Well, until the late 1960s, the glow-in-the-dark faces of clocks and car and aeroplane dashboard instruments glowed because they were painted with radium. So he started the slow process of haunting junk and antique shops, surreptitiously chipping off the glowing radium. But one day, he got lucky when his geiger counter went off its brain near an old clock. He bought the clock for $20, and inside, found a complete vial of radium paint which had been conveniently left behind.

So he rigged up a more powerful neutron gun with a hollow lead block. In a hole in the block he placed his precious radium and some beryllium which would be hit by the α-particles and give off neutrons. What did he use for a target? Some uranium ore purchased from a friendly geological supply shop. But failure again. The neutrons were moving too fast (at about 27 million kph) and just zipped through the uranium. So he slowed them down to about 8,000 kph by running them through tritium (which he painstakingly scraped off modern glow-in-the-dark gun and bow sights) — and the uranium ore got more radioactive.

a ball of radioactivity

By this time, David Hahn was 17 years of age, and he decided to stop fooling around. He mixed his radium with his americium and aluminium, wrapped it in aluminium foil, and then wrapped the whole mess in his thorium and uranium — of course, all held together with gaffer tape. Finally he had success — the bizarre ball got more radioactive every day. Perhaps too successful — he could pick up the radioactivity five houses away. He panicked and began to dismantle his creation.

At 2.40 am on 31 August 1994, the local police were called because a young man was behaving suspiciously near a car. (This could easily happen to you, if you try to illegally dispose of high-level nuclear waste.) David told the police to be careful of his toolbox, because it was radioactive. Soon some men in ventilated white moon-suits were chopping up his radioactive shed with chainsaws, and stuffing the parts into 39 sealed 200-litre drums which were taken to a nuclear waste repository. The clean-up cost about $120,000 — but it did protect the 40,000 nearby inhabitants from harm.

And David? Well, while he was a whiz at science, he never was much good with maths and English. So today, he's a junior sailor/deckhand on the aircraft carrier, USS *Enterprise*, which has eight nuclear reactors.

And if George Bush ever needs to call in the heavy artillery, perhaps he should forget the SEALs and the SAS, and call in the boy scouts to do their bob-a-job.

Nuclear University

Today, America has about 10,000 nuclear warheads. Six thousand of these are big strategic megaton nukes, designed to destroy military facilities or cities anywhere in the world. Another 1,300 are smaller tactical nukes, to be used on a battlefield. The remaining 2,700 are sleeping in storage.

There are two Big Problems.

First, the American military don't know how many of the 10,000 nukes still work. Most American nukes in their arsenal have an expected lifetime of 10 years — yet their average age is 20 years. The United States made their last nuclear weapons back in 1992.

The second problem is that they have forgotten how to make them. By 2007, half of the nuclear weapons designers now at Los Alamos will have retired — and they will all have retired by the year 2014. America has 10 times more NASA astronauts than nuclear weapons designers.

Most of America's nuclear weapons builders (going back to the early 1940s) are dead or retired. The knowledge of how to build a nuke exists only in the brains of a small number of people.

Why didn't they write down their hard-won knowledge?

First, because practically everything to do with nuclear weapons was, and still is, classified. If they wrote anything down, the engineers would then have their workload massively increased by having to do all the paperwork to classify and keep their documents secret.

Second, there was such a rush back in the Cold War to build new and better nukes, that they were too busy to write down exactly what they had done. A typical comment would be: *'I have an awful lot of stuff in my head that I've never bothered to write down. Part of that is just laziness, and part of it was that it was always more fun to move onto the next thing'*.

This is why the United States government has fired up The Nuclear University at its Los Alamos Laboratory. Their second class (of 11 students) graduated at the end of 2002. It's officially called TITANS — Theoretical Institute for Thermonuclear and Nuclear Studies. TITANS is one of the most exclusive educational institutions in the entire known Universe — and probably the only place you can get a degree in thermonuclear weapons.

The three entry qualifications are pretty demanding. First, you need a PhD in Chemistry, Physics or Engineering. Second, you need an ultra-high Level Q Security Clearance, and finally you have to be employed in X Division, the elite nuclear weapons design department at Los Alamos. The first two years in the classroom are spent studying everything to do with nuclear weapons — their history, design, physics, computational modelling, hydrodynamics and even particle transport. The final year is spent on an individual project, such as the W80 nuclear warhead, which is designed to go onto cruise missiles.

The few remaining learned types are teaching the new generation of scientists/engineers how to make bombs. They are like a parent teaching a child how to ride a bicycle with a mixture of theory, practice and verbal advice from the footpath.

But one thing's for sure, you won't get the recipe for making a nuke off the net — you'll have to go to Nuclear University. I wonder what pranks they get up to on Muck-Up Day?

References

Silverstein, Ken, 'Tale of the radioactive boy scout', *Reader's Digest*, March 1999, pp. 81–87.

When it comes to poisonous creatures, Australia is a world leader. And that includes some of our trees.

stinging tree

Australia is supposedly home to the nine most poisonous spiders in the world. It is true that we do have the most venomous snake in the world — the Fierce Snake. And it's also true that we have the world's most painful plant — the aptly-named Stinging Tree.

In fact, there are six species of Stinging Tree in Australia, but only two of them are the tall woody species — the other four are shrubs. They are rainforest plants and live along the east coast of Australia, from Cape York in the north of Queensland to Victoria in the south. They grow only where there is both strong sunlight and protection from the wind. So you'll see them along tracks, on the banks of creeks, and where the rainforest canopy has been broken by a falling branch or tree. They also pop up after a storm has ripped through a forest, or in areas cleared for development.

You sting me to tears!

Stinging Trees play an important part in the ecology of a rainforest. Many native Australian animals, birds and insects eat the leaves and fruit (they're not bothered by the sting).

stinging trees are vicious

We know a lot more about Stinging Trees, and what lives off them, thanks to Dr Marina Hurley, an entomologist and ecologist. She received her doctorate at James Cook University in Townsville, for her research into two shrubs, the Gympie-Gympie *Dendrocnide moroides* which is the most painful of all the Stinging Trees, and another closely related but slightly less harmful shrub, *D. cordifolia.* Both have large heart-shaped leaves and berry fruit.

Even though they don't hunt in packs, these Stinging Trees are pretty vicious. The sting is delivered through tiny silicon hairs that cover the leaves and the fruit of the plant. Think of these silicon hairs as tiny fibres of non-transparent glass. Dr Hurley found that the only way she could handle the leaves safely without getting stung, was to wear incredibly thick and bulky welding gloves. These silicon hairs penetrate your skin and then break off. They are so tiny, that often the skin will close over the hairs. So sometimes, once you've been stung, you can't remove the stinging hairs.

The silicon hairs cause pain, because they carry a neurotoxin. One scientist, Oelrichs, who purified the poison, injected himself with it and suffered intense pain, so proving that the toxin, not the silicon hairs, caused the pain. Once you have been stung, heating or cooling your skin can release even more neurotoxin. This neurotoxin is very stable. Leaves collected nearly a century ago can still cause pain.

The reaction varies, depending on the species of animal being stung, and how much of the skin gets stung. But humans feel something between mild irritation and intense pain and death. The pain strikes immediately after touching the plant, gradually increasing to a peak after about 20–30 minutes. The Dutch botanist, H.J. Winkler recorded the only official 'death by Stinging Tree', for a human — in New Guinea, in the early 1920s. There have been other anecdotal stories from soldiers in World War II suffering intense pain, and of an officer shooting himself because of the unrelenting pain.

But you can suffer pain even if you don't touch the plant. Dr Hurley discovered that the plants continuously shed their stinging hairs. Stay close to the Stinging Trees for more than an hour, and you can get an allergic reaction — intense pain and continuous bouts of sneezing. You can even get nose bleeds from silicon hairs floating in the air. But Dr Hurley found that if she wore a filter mask, which she replaced regularly, she could work near the plants for a few hours at a time.

Lovely Stinging Trees

Stinging trees are important in the ecology of the rainforest.

First, they fill in areas where the original cover has been broken, such as by death of ageing trees.

Second, their leaves are rich in nitrogen and calcium. So a shy, fleet-footed rainforest wallaby (the red-legged pademelon) loves to eat the leaves of the *D. moroides* Stinging Tree. The trees respond by being able to grow new leaves rapidly, and being able to quickly colonise any gaps that appear in the rainforest.

weird stingers

There are two weird things about these Stinging Trees. First, they are harmless to many native Australian species of fauna, but very nasty to introduced species such as humans, horses and dogs.

The second thing is even weirder. The pain is real and intense, but your body does not suffer any damage. Fire and snake bites cause pain *and* damage you as well. But it seems that the pain from this tree could be the only pain that is not related to any damage. Your lymph nodes swell up painfully and your skin feels an intense burning sensation but after it has worn off there is no damage to the skin. If we look at the neurotoxin involved, we might learn a lot about the mechanism of pain.

So what's the best way to remove the hairs, once you've accidentally been stuck on a Stinging Tree? Don't be tempted by the well-known 'cures', such as rubbing the affected area with the sap of a nearby elephant ear plant, or even with the ground-up roots of the tree that stung you. A student from James Cook University in Townsville discovered the best way to remove these hairs was with a non-heatable *Mariana* hair-removal wax strip. In fact this is now the official recommendation in the Queensland ambulance manual.

It might sound like a good way to get a free leg wax, but the pain is not worth it.

So next time you're working in the forest stick to the well-used tracks and if nature calls, be careful what you use as your toilet paper.

References

Hurley, Marina, 'Selective stingers', *ECOS*, vol. 105, October–December 2000, pp. 18–23.

We all love dunking our favourite bicky into a nice hot cuppa tea. But is this the best liquid for biscuit dunking?

biscuit dunking

Biscuits (or what the Americans call 'cookies') have been around for about 2,300 years, but only very recently have we started applying science to biscuits. I'm not talking about how to *make* the best biscuits, but the serious business of how to *dunk* biscuits. Thanks to physics, we have discovered which biscuits are best for dunking, and we have even discovered that tea and coffee are not the best liquids for biscuit dunking.

Biscuits appeared in Rome around the 3rd century BC. The word 'biscuit' comes from the Latin *'bis coctum'* which means 'twice-baked'. In those days, a biscuit was a thin unleavened wafer, which was quite hard and had a very low water content — hence the name 'twice-baked'. The advantage of the low water content was that the biscuit would have a long shelf life, because it wouldn't become mouldy. This, of course, was a few thousand years before tea and coffee made it to Europe, so the ancient Romans would soften their hard biscuits by dunking them in wine.

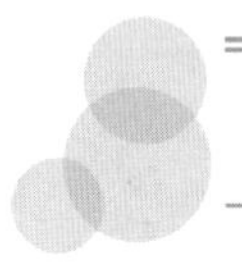

types of bickies for dunking

Nowadays the preferred liquids are hot tea or coffee. But there are many different factors to consider when dunking a biscuit. After all, you don't want your biscuit to disintegrate, leaving you with an unattractive sludge at the bottom of your hot cuppa. Some new-fangled modern biscuits have a central creamy section which is prone to melting, leaving behind two tasteless wafers, while other biscuits will simply collapse into a sloppy mess. Some biscuits are too stiff and rigid to enjoy easily *before* dunking, but are pleasantly edible *after* dunking. However, structural integrity is only part of the story — there's taste as well. For example some biscuits are boring and tasteless before you dunk them, but delicious after you dunk them.

So in 1998, Dr Len Fisher (working out of the University of Bristol in the UK) decided to look at the Physics of Biscuit Dunking.

A biscuit is basically dried-up grains of starch, which are glued together with sugar. When a biscuit is dunked, the hot liquid will

swell and soften the grains of starch — which is good. But the hot liquid will also *dissolve* the sugar, so that eventually the wetted biscuit loses so much structural integrity that it will collapse under its own weight. A dunked biscuit gets wet because it is porous. It is riddled with interconnecting hollow channels. Once the tea or coffee gets access to these channels, capillary action sucks the liquid deeper into the channels.

Len Fisher used an old equation from 1921 to predict how long it would take for the liquid to rise in your favourite biscuit. But he did more than just scribble equations — he did experiments, involving gold, a belt-sander, a microscope, an x-ray machine and sensitive weighing scales. He found that the best dunking time for a ginger nut biscuit was three seconds, but eight seconds for a digestive biscuit. Overall, his personal recommendation was to use a wide-brimmed cup filled almost to the top, to do horizontal dunking (so that only the bottom side of the biscuit gets wet), and then to quickly turn the dunked biscuit upside down so that the stronger dry side gives structural integrity to the wet side.

Animal Cookies

Biscuits (or cookies) in the shapes of animals appeared in the United Kingdom in the 1890s. But they became really popular during the Christmas of 1902 in the USA.

The National Biscuit Company of America (Nabisco) had released an edible toy for the Christmas market. Inside a little box of Animal Crackers were 22 little cookies (or biscuits). They were called 'Barnum's Animals' following on from the circus entrepreneur P.T. Barnum.

Nabisco produced 18 different animal shapes, including a walking and a sitting bear, a kangaroo, a gorilla and a zebra. But the boxes of animal crackers were filled randomly by machines, and so very rarely would a single box of 22 Animal Crackers have the complete complement of 18 different shapes.

This desire to collect, and then eat, all 18 different shapes gave an unexpected extra boost to the sales of Animal Crackers. Parents even wrote into the company with some important market research. They noted that American children would dismember the Animal Crackers in a specific order — starting with the back legs, then the front legs, then the head and finally the body.

it's all in the flavour

But in 1999, Len Fisher decided to revisit his experiments and look more deeply at *flavour*, rather than structural integrity. He wanted to work out which liquid was the best for dunking. To his surprise he found that it was not hot tea or coffee — but a milk drink.

There are two factors to flavour — taste (where the flavour chemicals excite the taste buds on your tongue) and smell (where the flavour chemicals excite the olfactory epithelium in your nose).

To analyse the smells, his team inserted a tube into one nostril of the eager volunteers. The different flavour chemicals from each sample of air were collected separated and then analysed. And sure enough, dunking your biscuit into a milky drink gives you up to 11 times more flavour release than from merely eating the dry biscuit.

Why? Well, the answer lies in the *fat* in milk. Milk is basically minuscule droplets of fat which are suspended in water. These droplets of fat do two things. First, they absorb the flavour molecules really well. Second, these little fat droplets also hang around in your mouth, so that the flavour and aroma chemicals can sit on your tongue *and* be released up to your nose.

But hot *non-milky* drinks tend to carry the flavour molecules straight down your gob and into your gut before the taste sensors on your tongue, and the smell sensors in the olfactory epithelium in your nose, have had a chance to appreciate them fully. His team also found that the worst drink to have with your biscuit is a soft drink such as lemonade. The flavour actually diminishes by a factor of 10. But these remarkable research results don't come easy — you've got to be prepared to work right through your morning tea break.

Koekje to Cookie

The modern sweet biscuit came from a small, sweet wedding cake from Holland called a *'koekje'*. The *'koekje'* is a diminutive of *'koek'*, which is the Dutch word meaning a full-size wedding cake. This small and very sweet *'koekje'* gave us the American word 'cookie'.

References

Fisher, Len, 'Physics takes the biscuit', *Nature*, vol. 397, 11 February 1999, p. 469.

Jensenius, Jens C., 'Many a slip "twixt cup and lip",' *Nature*, vol. 398, 18 March 1999, p. 186.

Nadis, Steve, 'Ig prizes spawn a new generation of Nobels', *Nature*, vol. 401, 7 October 1999, p. 518.

Panati, Charles, *Extraordinary Origins of Everyday Things*, Harper and Rowe, New York, 1987, pp. 411–415.

A third of a century after Neil Armstrong landed on the Moon, it has been confirmed that the Moon is suspiciously related to cheese.

moon and cheese

By an incredible cosmic coincidence, the impacts of Comet Shoemaker-Levy onto Jupiter happened exactly 25 years after the first landing of humans on the Moon. Even today, a third of a century since Neil Armstrong first set foot on the Moon, we are still discovering new things about our near neighbour, thanks to the experiments left behind — and one thing that we have proved, is that the Moon is suspiciously related to cheese!

Moon rocks are one of the richest sources of information about the Moon. The American astronauts brought back 382 kg of Moon rocks from six separate sites, while a remote-controlled Soviet craft brought back a few more kilograms. These rocks have answered many questions.

what was discovered from moon rocks

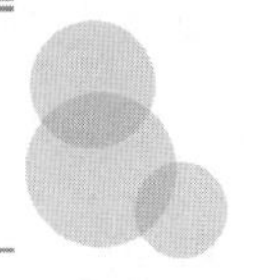

Before the Moon landings, we didn't know where the Moon came from. We once thought that the Moon might have been captured by Earth, or that it was spun off from Earth and flung into space, or perhaps that it was formed at the same time as Earth as part of a double-planet system. But now we think that the Moon was formed from the debris thrown into space, by the impact of a Mars-sized object smashing into Earth some 20 million years after it was formed. This debris then condensed into our Moon, which in those days, orbited much closer to Earth.

But besides the rocks we've brought back from the Moon, another important source of knowledge has been some special mirrors left behind on the surface of the Moon. These mirrors were actually clusters or arrays of corner-cube reflectors.

A corner-cube reflector has a very special property — it bounces back a beam of light in exactly the direction that it came from. You can see microscopic corner-cube reflectors at work at night, when your headlights swing across the number plate of another vehicle. There are tiny corner-cube reflectors buried in the paint of the number plate. A corner-cube reflector is like a box with one of the six sides removed. When a beam of light enters the box, regardless of the angle of entry, it will bounce off one of the sides and the back wall, and re-emerge at the same angle that it entered the box.

Three corner-cube reflector arrays were left on the Moon by the Americans from Apollo missions 11, 14 and 15, while another corner-cube reflector array is a French-built model installed on the Soviet Moon-roving vehicle, *Lunakhod 2*. Each of these four corner-cube reflector arrays is roughly the size of a large breakfast cereal packet.

La famille de fromage et de lune
(the cheese and moon family)

beams of light

American and French scientists use the mirrors left on the Moon to conduct lunar laser ranging measurements about three times a day, to measure the distance between Earth and the Moon. They shine pulses of laser light through a telescope, at one of these corner-cube reflectors. Athough the laser beam is only a few centimetres wide when it is fired out of the telescope on Earth, by the time it reaches the Moon, it has spread out to 7 km in diameter. So the corner-cube reflector array catches only one billionth of the incoming laser light.

The corner-cube reflector array then bounces the laser beam back to Earth. Once again, the light beam spreads, and it ends up as a spot, about 20 km in diameter, centred on the observatory. A telescope on the receiving end, about one metre in diameter, will see about one billionth of the returning photons of light. When you take into account that there are only about 10 billion billion photons in the initial outgoing laser pulse from Earth to the Moon, the receiving station on the ground has to be able to count individual photons of light — so it has to be a very high-technology piece of equipment.

The distance between Earth and the Moon varies because both Earth and the Moon do not orbit perfectly evenly but wobble a bit in their orbits, *and* the gravity of the Sun distorts the orbits as well. By taking measurements over a third of a century, they have been able to improve by a factor of thousands, their knowledge of the movement of the Moon around Earth, and the Moon's speed of rotation. They have been able to test, yet again, Einstein's Theory of General Relativity or Gravity, and they have even discovered a strange 50-day-long oscillation in the length of our Earth day, which is somehow linked to a similar oscillation in the atmosphere.

what we discovered about the moon

They have even been able to work out, by looking at the wobble of the Moon, that just like Earth, it is layered like a giant onion. It probably has a 60-km-thick crust which is about 2.75 times denser than water. Underneath the crust are two separate layers. Finally at the centre, there is an iron core, about 200–400 km in diameter, with a density seven times that of water. This core could actually be liquid iron, with a turbulent boundary layer.

A lot of this information was worked out just recently. But because these corner-cube reflectors are entirely passive and need no electricity, they'll keep on providing scientific data for many years to come.

the moon is a type of cheese

There is one scientific result, however, that the scientists are not publicising — and it's all to do with cheese! You may remember that many children's fairy tales claim that the Moon is made of some type of cheese. Strangely, this has actually been backed up by seismic data transmitted from the Moon.

They measured the speed of sound in the Moon rocks and calculated it to be somewhere between 1.2 and 1.8 km per second, depending on what type of Moon rock was being looked at. This is much lower than in Earth rocks — where the speed of sound is around 5–6 km per second. It so happens that the speed of sound in cheeses such as Romano and cheddar is around 1.5–1.8 km per second — the same range as the speed of sound in Moon rocks!

The scientists claim that the *speed of sound in cheese* being very close to the *speed of sound in Moon rocks* is just a coincidence, and quite easy to explain. They say that there's no water in Moon rocks, and Moon rocks are very porous with lots of holes. These two factors make sound travel very slowly in Moon rocks, as compared to Earth rocks.

But perhaps the scientists' explanations are just part of a giant conspiracy theory, and those of us who believe in fairy stories and magic can rest safe in the knowledge that the Moon really is made of cheese.

Moon and Cheese Raw Data

SEISMIC VELOCITIES IN LUNAR ROCKS, TERRESTRIAL ROCKS AND SELECTED CHEESES*

CHEESES	VELOCITY IN KM/SEC
Sapsego (Swiss)	2.12
Romano (Italy)	1.74
Cheddar (Vermont)	1.72
Muenster (Wisconsin)	1.57
LUNAR ROCKS	**VELOCITY IN KM/SEC**
Basalt 10017	1.84
Basalt 10046	1.25
Near-surface layer	1.20
TERRESTRIAL ROCKS	
Granite	5.9
Gneiss	4.9
Basalt	5.8
Sandstone	4.9

*Source: Schreiber and Anderson (1970)

So it is obvious that the velocity of sound in cheese is very close to the velocity of sound in Moon rocks — and quite different from the velocity of sound in Earth rocks.

References

Dickey, J.O., *et al.*, 'Lunar laser ranging: a continuing legacy of the Apollo Program', *Science*, vol. 265, 22 July 1994, pp. 453, 482–490.

Ringwood, A.E., 'Origin of the Earth and Moon', *The Australian Physicist*, vol. 18, June 1981, pp. 91–102.

Runcorn, S.K., 'The Moon's ancient magnetism', *Scientific American*, December 1987, pp. 34–42.

Taylor, G. Jeffrey, 'The scientific legacy of Apollo', *Scientific American*, July 1994, pp. 26–33.

DNA is the blueprint of a living creature. But one set of DNA makes two very different creatures.

wasp makes virus

As we get deeper into the genetic revolution, you will hear more and more about DNA. In fact, you've probably already heard somebody say that DNA is the blueprint of a living creature — that it has all of the instructions needed to make a living creature. But would you believe that one set of DNA can make two very different living creatures!?

It sounds really odd, so let me give you an example. As you know, sheep farmers work very closely with their sheep dogs. You also know that humans are made by human DNA, and sheep dogs are made by sheep dog DNA. But what if the human DNA made both the human and the sheep dog!? Well, this is what happens in one type of wasp according to research done by Nancy E. Beckage, a Professor of Entomology at the University of California at Riverside.

wasp injects the caterpillar

You might have heard that some wasps inject their eggs into the body of a caterpillar. Most babies need concentrated high-energy foods that are packed with fats, proteins, carbohydrates and vitamins. The body of the caterpillar provides these very goodies for the wasp eggs. At the same time as injecting the eggs, the mother wasp injects a liquid which contains a virus. This virus behaves a bit like the AIDS virus, severely damaging the immune system of the caterpillar. Thanks to the virus, the caterpillar now has a very weak immune system, and cannot kill the eggs. After a fortnight or so, the wasp eggs have grown into baby wasps, and the caterpillar finally dies. You can see that the virus is really essential to help the wasp eggs use the caterpillar as a convenient food supply.

If the liquid (which contains the virus) is washed away from the eggs, and the naked eggs (without the virus) are injected into the body of the caterpillar, the immune system of the caterpillar will soon fight them off successfully. But if the washed eggs (which are free of the virus) are injected into the caterpillar at the same time as a separate injection of the virus particles, the caterpillar loses the fight, and the wasp eggs win.

Within 30 minutes of being injected, the virus begins to spread throughout the caterpillar, entering many of its cells. The immune response of the caterpillar does return to normal levels within a few days, but by then, it's too late for the caterpillar.

Now here's the really weird bit. The DNA that makes the virus is *not separate* from the DNA that makes the wasp. In fact, the DNA that makes the virus is jumbled up inside the DNA for the wasp! So are the wasp and the virus two separate living creatures in a very tight relationship, or are they one living creature?

The early wasp gets the caterpillar host!

When one life ends, another begins...

And the 2 Become 1 . . .

According to biologists, living creatures tend to have four main types of close partnerships.

In an *epizoic* relationship, one of the partners lives on the skin of the other one. A well-known example is the remora fish, which will attach itself to a shark thanks to a suction cap on the top of the remora fish's head.

A *commensal* relationship is a one-sided relationship where only one of the partners benefits — but it doesn't cost the other partner anything at all. So the bird called the cattle egret eats the parasites from the backs of grazing animals such as cattle, but doesn't harm these grazing animals.

The relationship between the caterpillar and the wasp can be called a *parasitic* relationship. In this one-sided relationship, the wasp benefits, but the caterpillar dies.

The fourth type of relationship is called *symbiotic*, which means 'living together'. A classic example of this relationship is that of a lichen. A lichen is actually two separate living creatures — it is a fungus and a plant called an algae. But in lichen, the fungus and the algae have quite distinct and separate DNA. Maybe a new word will have to be invented to describe the relationship between the wasp and the virus that it lives with.

wasp and virus wins

How did this incredibly close relationship between the wasp and the virus come about? Certainly, they both benefit. The *babies of the wasp* benefit, because they get a nice convenient food supply, ensuring that the line of wasps continues for generation after generation. And the *virus* benefits. The virus spreads throughout the entire body of the caterpillar, infecting all its cells, and at the same time, making more viruses. So both the wasp and the virus come out winners.

Was the virus originally a separate creature that reproduced inside the caterpillar, and, much later, got incorporated into the wasp's DNA? Or perhaps the viruses never were a separate creature? Perhaps the wasps evolved the ability to modify their own wasp DNA so that it would make a virus that would immobilise the caterpillar, and keep it as a convenient food supply for the wasp eggs. At this stage, we simply don't know.

In North America, jewel wasps use cockroaches as the preferred home for their eggs. After the eggs have been injected, the cockroach is kept alive for a few weeks before the larvae pupate, and emerge from the body of the cockroach as jewel wasps. Of course, to put it delicately, the cockroach is expended.

If we can just genetically engineer the wasp, we might have finally discovered how to get rid of cockroaches!

References

Beckage, Nancy E., 'The parasitic wasp's secret weapon', *Scientific American*, November 1997, pp. 50–55.

Exploring The Secrets Of Nature, Reader's Digest Association Far East Limited, 1994, pp. 23–24, 57, 64, 124.

Furlow, Bryant, 'The Enemy Within', *New Scientist*, 19 August 2000, pp. 38–41.

Miller, Robert V., 'Bacterial Gene Swapping in Nature', *Scientific American*, January 1998, pp. 46–51.

Tumlinson, James H., Lewis, W. Joe & Vet, Louise E. M., 'How parasitic wasps find their hosts', *Scientific American*, vol. 266, no. 3, March 1993, pp. 46–52.

The ancient game of Rock–Paper–Scissors does much more than just settle a dispute between two people — this game can work out which of the nerves that control your hand are damaged.

rock-paper-scissors

Your hand is that incredibly important prehensile (or grasping) organ at the far end of the multi-jointed lever called the 'upper limb' (the rest of us call it an 'arm').

What makes the hand exceptional is the thumb, which is probably more important than the other four fingers of the hand. If you activate one set of muscles, your hand can form a 'static hook', so that you can grab a briefcase. Another set of muscles gives you the 'pinch grip', where the index finger and thumb combine to let you make precision movements, such as passing a thread through the eye of a needle. Yet another set of muscles lets you make the 'power grip', which allows you to grab the handle of a hammer.

Nerves control all of these different muscles. And medical doctors have to know which nerves control which muscles of the hand, and that's where a simple game can assist their memory.

nerves control hand

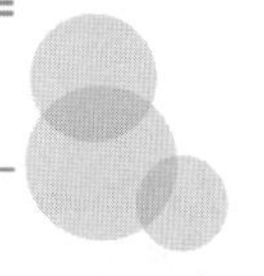

The hand is a remote-powered and remote-controlled device.

One strange fact about the hand is that practically all the muscles that make it do stuff are in the forearm, near the elbow. The muscles pull on the fingers via long tendons.

In the same way, an important stage in the control of the hand doesn't happen locally, but in the spine of the neck.

Here, five nerves (C5, C6, C7, C8 and T1) run out from the four lowest cervical vertebrae and the first thoracic vertebra (T1). If each of the five nerves ran one of the five fingers of the hand, it would be easy to remember. But if you damaged one of these nerves, you would lose all function to the finger it controlled.

So in an area near your shoulder called the 'brachial plexus', the five nerves all mix-and-match and get all crossed over. They arrange themselves into three nerves that control the hand — the median nerve, the radial nerve and the ulnar nerve.

Here's an example of this mix-and-match cross-over: the ulnar nerve is made up of C8 and T1 (and sometimes, from C7), but no nerves from C5 and C6. This crossing-over gives some protection in case of injury — a finger or two might get weaker, but probably won't go totally floppy. For instance, if you damage the C5 and C6 nerves, it would not affect the ulnar nerve and the muscles it controlled.

The arrangement of these nerves and the muscles they control is quite complicated, and memorising them has worried medical students for the last century. As junior doctors are often the first to see patients with injuries to the arm, it's important for them to know which nerves control which muscles of the hand.

These injuries can sometimes be treated on a spectrum ranging from watching and waiting to see if the body heals itself, to immediate and invasive surgery — and it's up to the junior doctor to make that important decision.

1. Never rely on Rock-Paper-Scissors to determine a life-threatening decision.

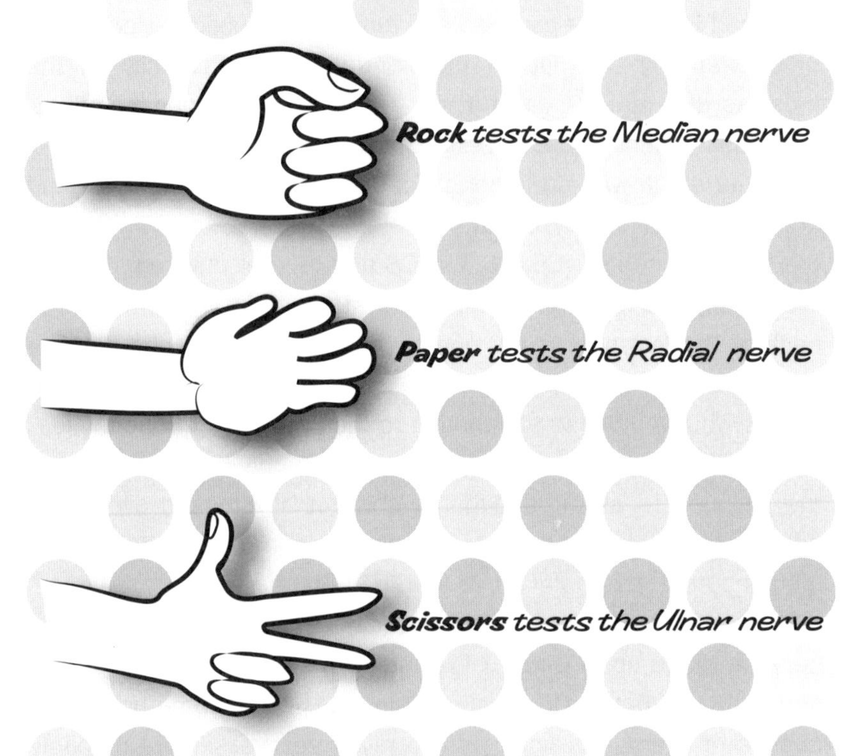

2. The hand is controlled by a complex wiring system of nerves. Rock-Paper-Scissors teaches you the nerves that control the muscles that move the fingers and thumb.

3. When playing Rock-Paper-Scissors, it's always a good idea to remove all hand jewellery.

You have to hand it to them...

knowing nerves

Dr A.W. Davidson, from the Department of Trauma and Orthopaedics at the Royal London Hospital, reckons that junior medical doctors don't really know the nerves and muscles of the hand. He gave a series of simple clinical questions about the three nerves that run the hand (median, radial and ulnar nerves) to 20 junior doctors. They could get any mark between 0 and 10. The average mark was three, and no doctor got a mark higher than five. So Davidson wrote a paper in the surgical journal *Injury*, entitled *'Rock–Paper–Scissors'*.

Saturday-Night Palsy

'Palsy' means a 'loss of function', and 'Saturday Night' is when this particular 'dysfunction' often occurs.

You usually move around in your sleep, so that you don't put pressure on one small area for too long. But when a person is under the influence of narcotics or too much alcohol, they lose this normal protective reflex. They are effectively temporarily paralysed. Occasionally, people will fall unconscious into a profound sleep with their arm over the back of a chair. This puts pressure on, and then causes damage to, the median and radial nerves. So a patient with Saturday-Night Palsy can't make 'rock' or 'paper', but they can still make 'scissors'.

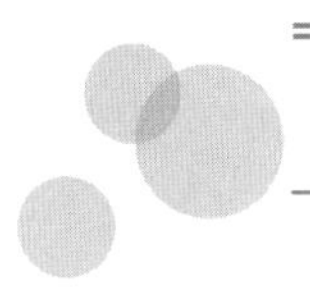

rock–paper–scissors

Rock–Paper–Scissors is a very ancient game. Back before scissors were invented, there was a similar game called 'Earwig–Man–Elephant'. (The earwig overcomes the elephant by crawling into its ear.) Today, Rock–Paper–Scissors is often used to decide matters between two people in much the same way they might toss a coin. On the count of three, each player has to select a hand position.

Now **rock** (which is a clenched fist) will break scissors. (Some purists say that it blunts the scissors.) But **scissors** (which has the index and second fingers open and outstretched, and the little and ring fingers tucked in) will cut paper. And **paper** (the open hand held flat with the fingers together and outstretched) will smother the rock.

So rock will break scissors, scissors will cut paper, and paper will cover the rock — and around and around it goes. You can win, lose or make a draw.

games and nerves

Now getting back to the hand itself, you can work out most injuries to the nerves that control the muscles to the hand by getting the patient to make a rock, paper or scissors.

The median nerve clenches all of the fingers, which gives you the 'rock' position. The radial nerve will extend or stretch out fingers from the closed position to the open position, so that gives you the 'paper' position. And the ulnar nerve does two things. First it makes the little finger and the ring finger next to it, tuck in. And second, it spreads the index and middle finger. So overall, your hand looks like a pair of scissors with the ring and little fingers clawed up and the index and middle fingers opening and closing. By using this simple game, junior doctors will be able to make a decent fist of diagnosing hand injuries.

Earwig has Spare Penis

The only reason I mention anything to do with 'earwigs' is that the game Earwig–Man–Elephant came before Rock–Paper–Scissors.

Earwigs have two penises, because one of them can break off. Dr Kamimura, from Tokyo Metropolitan University, wrote about this in the scientific journal, *Naturwissenschaften*. In biology-speak, he says that *'males have paired, elongated male intromittent organs'*.

The name 'earwig' comes from the Anglo-Saxon word for 'ear creature'. It was widely believed these creatures would crawl into the ears of sleeping people. There are about 1,100 species of earwigs, ranging from 5–50 mm long. They all have large membranous hindwings under their short forewings. They are nocturnal, and usually vegetarian.

Scientists had long assumed that one of the two penises could not work, because it pointed in the 'wrong' direction. (But, as your teacher said in kindergarten, 'assume' makes an 'ass' of 'u' and 'me'.)

Dr Kamimura didn't assume. He went to the trouble of looking at twelve earwigs from the family Anisolabididae having sex. Earwigs from this family have exceptionally long penises — usually longer than their entire body. He then interrupted them

90 seconds after they had started. He found that 9 of the 12 earwigs had lost one of their penises — it had simply snapped off during sex. But within two days, they were able to successfully inseminate a female with the remaining penis. So he called his paper, *'A "spare" compensates for the risk of destruction of the elongated penis of earwigs'.*

The earwig has taken the Scouts' motto of 'Be Prepared' to the max.

World Rock Paper Scissors Society

The World RPS or Rock Paper Scissors Society was formed in 1918. On their homepage (www.worldrps.com), they give basic advice and etiquette. For example, they encourage Safe RPS Playing: *'Always ensure that all players have removed sharp jewellery and watches.'* They recommend that the 'paper' hand should be horizontal, and that *'use of the vertical paper (sometimes referred to as "the handshake") is considered exceptionally bad form'.*

They also make recommendations on the rules of the game, and advanced gambits. They reckon that you should not play random moves. You should either use psychology to anticipate your opponent's moves, or use a certain sequence of moves to psychologically influence your opponent's responses. For example, a subtle, yet aggressive gambit would be to throw three Rocks in a row — Rock–Rock–Rock. An invasive and devious gambit would be to throw the triple called the Scissor Sandwich — Paper–Scissors–Paper. And a tribute to those who often win in the end through sheer doggedness is the Bureaucrat — Paper–Paper–Paper.

To get you through very difficult times, they even offer Haiku (Japanese 3-line 17-syllable poems), such as this one:

'stone, scissors, paper
safety is an illusion
must fight, run, or hide.'

thumb up

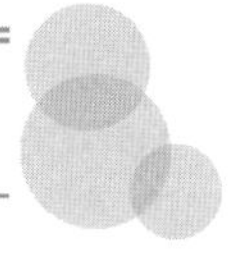

According to the World Rock Paper Scissors Society, 'scissors' is made with the thumb bent over to cover the curled fourth and fifth fingers. This leads to a problem.

If you want to bend your thumb over your curled fourth and fifth fingers, you have to use your median nerve. But this would mean that making a 'scissors' would no longer be under 'pure' ulnar nerve control.

So if you're a doctor, here's what you do if you want to use Rock–Paper–Scissors to test the specific action of the individual nerves that control the hand. Ask the patient to not curl the thumb over, but instead, leave the thumb pointing up. Some people throw a 'scissors' like this anyway.

funny bone

The ulnar nerve winds its way down the arm from the shoulder to the hand, and passes over the bones in the back of the elbow. It is relatively exposed here, and is occasionally bumped against something.

The bump causes a temporary paralysis, and a vaguely unpleasant 'tingling'. The damage is caused by the bump to the nerve, not the bone, but the nerve is hidden and the bone is obvious — hence the name, 'funny bone'.

Next time you bump your 'funny bone', see if you can still do a 'scissors' during that short five-minute paralysis (ignore the pain, and do it for 'science'). The ulnar nerve also controls the finer muscles of the hand, so at the same time check out your ability to do delicate fine work with your hand.

References

Davidson, A.W., 'Rock–paper–scissors', *Injury*, vol. 34, no. 1, January 2003, pp. 61–63.

Fannin, Penny, 'Ear's to a penis or two', *Sydney Morning Herald*, 3 December 2001, p. 3.

Gardener, Ernest Dean & Gray, Donald James, *Human Anatomy*, W.B. Saunders Company, 1975, pp. 109–114, 136–138, 142–150, 765–771.

Kamimura, Yoshitaka & Matsuo, Yoh, 'A "spare" compensates for the risk of destruction of the elongated penis of earwigs (Insecta: Dermaptera)', *Naturwissenschaften* (2001) 88:468–71.

Warwick, Roger & Williams, Peter L., *Gray's Anatomy*, Longman, 1973 (35th edition), pp. 1034–1047.

World RPS Webpage: http://www.worldrps.com

A 'sausage poison' has become one of the most powerful toxins known to mankind. It helps people walk, sleep through the night and relieve headaches. But it's really famous for making people 'beautiful' by helping fight the war against wrinkles.

beautiful botox

In 1793, in Wildbad, Germany, 13 people all ate some uncooked smoked sausage. They all got food poisoning, and six of them died. The Latin word for sausage is *'botulus'*, so the disease which killed them was called 'botulism'. These unlucky souls had eaten one of the most powerful toxins known to the human race.

bug makes poison

It took until 1896 before Émile van Ermengem discovered that this powerful poison was made by a bacterium. The bacterium is now called *Clostridium botulinum*. (To save time, microbiologists often shorten the first name of a bacterium to its first letter, so *Clostridium botulinum* becomes *C. botulinum*.)

The Clostridium family of bacteria is big, and most of them are unfriendly to humans. They live in the soil, or in water, or in the gut of humans and animals. These bacteria look like tiny rods — about half a micron (a micron is a millionth of a metre) across and about 5 microns long.

Clostridium tetani causes tetanus, while *Clostridium perfringens* causes gas gangrene. *C. perfringens* produces gas, which pushes its way **between** the layers of the skin, tearing blood vessels apart. The flesh of somebody afflicted with gas gangrene feels like bubble wrap — with the bubbles a few millimetres across.

And finally, there's *Clostridium botulinum* which causes botulism. The bacterium doesn't like air, so the inside of a smoked sausage or a can of food makes for it an ideal home.

three types of botulism

The classic type is **foodborne botulism**, which you get by eating food contaminated with a botulinum toxin. The toxin passes unchanged through the gut, enters the blood vessels lining the gut, and then travels via the bloodstream to the nerve endings in muscles. The symptoms often start with blurred or double vision, dry mouth and difficulty in swallowing, and then proceed to symmetrical descending floppy paralysis of the muscles. Frighteningly, the person can be completely mentally alert as their arms, and then their legs, go flaccid. An intravenous injection of botulinum antitoxin often helps. In wealthy countries, the death rate is around 5–10% — but even those who survive can still take months to recover. The symptoms of foodborne botulism usually appear within 12 to 36 hours. The sooner they appear, the more severe the disease and the higher the death rate.

Foodborne botulism is much less common today than it was

when people did their own home preserving — certainly a case where modern food technology has improved our lives.

The second type of botulism is **wound botulism**, which has the same clinical picture as foodborne botulism. The bacterium contaminates a wound that is kept away from the air.

But the most common form of botulism in wealthy countries is the third type — **intestinal** or **infant botulism**. It usually occurs in infants under a year old. The babies swallow the spores of the bacteria, which then grow in the large intestine and make toxins. The symptoms usually start with constipation, then move on to tiredness, poor feeding, droopy eyes, difficulty in swallowing and then a generalised floppiness. Intestinal or infant botulism may be involved in some 5% of cases of SIDS (Sudden Infant Death Syndrome). In wealthy countries, the death rate is less than 1%.

types of *c. botulinum*

There are at least seven different types of *C. botulinum* bacteria. Each one produces its own different version of nasty botulinum toxin. The toxin from Type A is called Botox — the name has been copyrighted by the Allergan company. Botulinum toxin Type B is sold as Myobloc or Neurobloc.

Most human foodborne botulism is caused by Types A, B, E and rarely, Type F. Most infant botulism is related to *C. botulinum* Types A or B. Type E botulism usually comes from eating seafood such as fish and meat from marine animals. Most bacteria go to sleep at low temperatures, but not Type E. The *C. botulinum* bacteria that make Type E toxin keep pumping it out at 3°C, which is lower than the temperature in your fridge.

C. botulinum Type G lives in the soil, and has been found at autopsy in humans. However, it has not been linked to causing botulism in humans.

bacteria, toxin and spore

The botulism toxin we currently know most about is *C. botulinum* Type A, and the toxin it makes (Botulinum Toxin Type A). The *C. botulinum* bacterium that makes Botox is quite fragile — 10 minutes at 80°C will kill it.

Botox toxin is even more fragile — two minutes at 70°C will destroy it. But it's very dangerous. This toxin is so powerful that as little as 100 billionths of a gram can kill. (It's one of the most lethal poisons known.) In fact, so little is required for medical uses, that in 1995, doctors around the world were still using some of the first medical (not weapons) batch made 18 years earlier in 1977!

But the spore (a robust 'sleeping' bacteria that can slumber for a century or more) of *C. botulinum* is very tough and hard to kill. It can survive several hours of boiling at 100°C. You need moist heat at 120°C to be sure of killing it. You can easily achieve this in a commercial food-canning factory. But if you are doing some home food canning, moist heat at 120°C is hard to achieve — you would have to use an industrial-grade pressure cooker. This spore is so tough that it can also survive most detergents and toxic chemicals, as well as being dried out.

The spore can survive for years and travel great distances on the wind. It can then wake up and start making toxin that paralyses your nerves.

terminator toxin

The toxin kills you by stopping nerves from talking to muscles. When your lung muscles stop working, you die.

When you decide to move a muscle, your brain tells your nerves to do it. For instance, the nerves tell the muscles of your right hand to turn the next page of this book. The nerves do this by squirting out tiny balls filled with a chemical called 'acetylcholine'. Thousands of molecules of acetylcholine travel across a tiny gap (about 20 billionths of a metre) and land on a muscle cell. So acetylcholine says to the muscles, '*contract now*', and then your hand turns the page.

Botox destroys the tiny balls of acetylcholine, as they sit inside the nerve where it comes close to a muscle cell. Once the toxin has bonded to the nerve endings, they can never release acetlycholine. Any muscles controlled by those nerves are permanently paralysed, because the toxin will never let go of those nerve endings. Our current medical science can't kick the toxin off the nerves.

However, after a few days, the nerve will start to grow tiny new nerve endings. These new nerve endings can release acetylcholine. So, over the next few months, you will eventually recover your full normal muscle control, that is, if you can get over the short-term paralysis of your lungs, gut, etc.! For example, the doctors might cut a hole into the throat (a tracheostomy), and install an air pipe. They then connect an artificial respirator to pump air in and out of the lungs, for as long as it takes, which could be days or weeks.

By the way, the toxin interferes only with the nerves which control the muscles. So, if you are poisoned with a botulinum toxin, your sensations are still normal — you can still feel heat, pain, etc.

So why do people inject this nasty botulinum toxin into their faces? (In fact, they love doing it so much that Botox injection is one of the most popular cosmetic surgery techniques of all time.)

Trojan Horse

Botulinum Toxin A works like a Trojan Horse.

Thirty-two centuries ago, the Greeks fought the Trojans for 10 years. The Greeks won victory by pretending to abandon their camp in defeat. They left behind a huge wooden statue, supposedly as a gift to their victors, the Trojans.

Legend has it that the Trojans rolled the horse into their fortified camp, and then spent the night getting thoroughly wasted. The Greeks, who were hidden inside the horse, jumped out and opened the gates to their Greek colleagues who had secretly returned — and that's how the Greeks won the Trojan War. This war gave us the phrase 'Trojan Horse' (meaning subversion from within).

Botulinum Toxin A is made up of 1,285 amino acids, which are arranged into three parts, each having a separate job.

The first part sticks to the cell membrane of the nerve cell, and then is swallowed up inside the cell. But it's still wrapped up in the tiny pocket of the cell membrane. The pocket now looks like a little ball.

The second part of the toxin pushes through the wall of this ball of nerve cell membrane. Most of the second part of the toxin remains stuck inside the ball.

A small active part of the toxin finally gets into the body of the nerve cell — this is the third part. This third part does the dirty work. It digests the little bubbles in the nerve cell that contain the chemical called acetylcholine. If the nerve cell no longer has any acetylcholine to release, then it can't talk to the muscles and the muscles become paralysed.

The Trojan Horse aspect to botulinum toxin is that the second part of the toxin loops around the third part like a belt covering a knife. So this ultra-dangerous third part (that destroys the acetylcholine) is hidden until the last moment.

This information tells us that there's no point in trying to design an anti-toxin to fight the third part of the toxin, because it's hidden from sight. Our biochemists would be better off developing anti-toxins that fight the stick-to-the surface-of-the-cell activity, or the push-through-the-tiny-pocket-of-cell-membrane activity.

search for beauty

About 8,000 years ago, our ancestors ground and mixed eye-paint and face-powder. In Ancient Egypt, about 6,000 years ago, the Egyptians made perfume and sold it in shops along with other beauty products. Back then, green was the cool colour for eye shadow. It was made from a green copper ore called malachite. For a big night out, a woman would tip her nipples with gold and paint the veins on her breasts blue. The most popular lipstick colour of that time was blue-black.

Around 5,000 years ago, the Chinese made a fingernail polish by mixing gum arabic, egg white, beeswax and gelatin. Fashions for the elite changed with time — the Chou Dynasty of 600 BC favoured gold and silver for fingernail polish, but 2,000 years later the Ming Dynasty preferred black and red.

People have traditionally suffered for their beauty. Eighteenth-century wealthy Europeans admired a pale, white face. Arsenic

Botox – the friendly Weapon of Mass Destruction

Fort Detrick, Maryland USA,
the 'home' of Botulinum Toxin ... or sausage poison

The Botox target areas: forehead creases, eyes, crow's feet, the corners of the mouth and ... common sense.

In 2000, 210,000 Americans had tried medical procedures with Botox. By 2001, this figure had rocketed to 1.6 million.

In 2002, the FDA finally approved Botox for wrinkle removal.

Complexion Wafers were very popular because they really worked (no false advertising). The arsenic poisoned the red blood cells, so there was less oxygen in circulation, making the skin deathly white.

In 2002, 7.5 million Americans chased beauty with everything from face-lifts and brow-lifts, to collagen injections. Gone are the days when a loving spouse would give a Christmas present of a few days at a health spa. Today, there are vouchers for Botox injections and liposuction under the Christmas tree. (Is that why Santa's holding his age so well? Is he knocking off the Botox vouchers?) Dr Philip Miller, a New York plastic surgeon, now offers gift certificates for wrinkle removal, ranging between US$500 and US$4,000.

Botox follows in the ancient tradition of trying to stop Time's Winged Chariot by buying temporary beauty.

warfare to wonderful

Like many other scientific developments in the world, Botox began with the military.

Around the middle of the 20th century, the US Army was interested in biological warfare. At Fort Detrick, Maryland, they were the first to develop a technique to purify botulinum toxin.

In 1980, Dr Alan Scott, an ophthalmologist, described one of the first medical uses for botulinum toxin. Normally, you swivel your eyeball up-and-down and left-to-right by contracting tiny muscles in your eye socket. These muscles pull on the globe of your eyeball, to aim it at something. But sometimes, if one of those muscles is over-active, it pulls one eye permanently to one side. Dr Scott described how he could bring the eyes parallel again by injecting botulinum toxin into the over-active muscle to paralyse it.

The 'beauty' effects of the 'sausage poison' were discovered in a happy accident. In 1987, Dr Jean Carruthers, an eye doctor at the University of British Columbia, was using the toxin on her patients to paralyse and weaken spastic eye muscles. Her husband, Arthur, was a dermatologist. They both noticed that after the injections, Dr Carruthers' patients had developed very smooth and unfurrowed brows. The plastic surgeons and the dermatologists started using it on the forehead, eyes and even the neck.

Botox is the One?

The name Botox is copyrighted. Botox (**Bo**tulinum **tox**in A) was the first botulinum toxin on the market, but there are others. The British company Ipsen makes Dysport — their version of botulinum toxin A.

Elan Pharmaceuticals markets Myobloc (botulinum toxin B). Myobloc is faster-acting, as it starts working within hours, whereas Botox takes several days to reach its full effect. On the other hand, at its currently used doses, Myobloc does not last as long as Botox.

beauty through a needle

Botox is usually injected through a very small needle, in a large number of small doses. Its effects begin after about two days, and reach full potency after three or four days. It smoothes out the small muscles on the corners of the eyes, the forehead and the area between the eyebrows. It leaves the patient with more youthful-looking eyes and a smooth forehead. But it does not make the skin feel numb.

This poison can sometimes get rid of the infamous 'worry lines' between the eyes, or the 'crows' feet' at the corners of the eyes, or indeed, wrinkles anywhere on the upper part of the face, by paralysing the relevant muscles. Mind you, it takes a few days for the toxin to work, and you have to repeat the treatment every three to six months, as it wears off.

After the injections, patients are advised not to move around too much for six hours, in case the toxin seeps into nearby areas and paralyses other muscles. So they can't bend over to tie their shoelaces, lie down, or even lean to one side for six hours.

There is currently no way to reverse Botox's effect, so you have to wait three to six months for it to wear off.

menace to magnificence

The Botox Boom had started.

In 1989, Botox was officially approved, but only as an 'orphan' drug, for tightly clenched eyelids and crossed eyes, and very spastic neck muscles. In 1991, the American FDA officially approved Botox as a treatment for eye muscle spasms. That same year, Allergan bought the US Army's entire supply of toxin, and in 1997 started making their own.

In 2000, 1.1 million Americans had tried cosmetic medical procedures — with Botox (at 210,000) being the most popular. In 2001, Botox use by Americans rocketed to 1.6 million. In 2002, 7.5 million Americans tried cosmetic procedures and Botox use escalated again.

Doctors were using Botox to remove wrinkles, but the manufacturer, Allergan, couldn't advertise this to the public. This changed in April 2002, when the American FDA approved Botox for wrinkle removal. Allergan's worldwide Botox sales were US$440 million in 2002, as compared to US$310 million in 2001, and US$19.5 million in 1992.

In the USA, a practitioner can buy a vial of botulinum toxin for US$400, then charge US$1,000 for injecting it. He can then repeat the process again with another patient 10 minutes later. It's extremely lucrative.

menace to medicine: muscles

Botulinum Toxin Type A has several hundred other medical uses aside from 'beauty' treatments, and the list is growing continually. As I said earlier, it's great for paralysing muscles. The paralysis is total or partial, depending on how much you inject.

In the condition called 'blepharospasm', the muscles of your eyelids can go hyperactive and contract so frequently and forcefully that you are virtually blind. Here, eye doctors can delicately paralyse these eyelid muscles, with a success rate better than 90%. The side effects are usually mild, such as dry eyes, tearing eyes, or local pain and swelling. The injections still work after seven years in practically all patients.

In 'torticollis', some of the muscles of your neck are overactive. Your head may be tilted to one side, or rotated, or tilted up or down, or you might even have a lifted shoulder. Once again, a selective and accurate paralysis of the relevant muscles can help about 75% of sufferers. The main side effect is difficulty in swallowing, but you can also suffer local muscle weakness and tiredness. About 5% of patients find the injections no longer work after a few years. (Maybe some of the other botulinum toxin Types B to G might work?)

People who have suffered a stroke often find their arm muscles go into spasm, leaving their arms permanently bent. Botulinum toxin Type A has had some success with some stroke patients.

Sometimes people can suffer from problems with muscles controlling the larynx. As they talk in a strained voice, the pitch can change unpredictably, and they can suddenly stop making sound. They often sound like they're being strangled whenever they try to talk. The selective paralysis of the botulinum toxin has helped this condition, as well as writer's cramp, musician's cramp, and the infamous 'golfer's yips' (where golfers can no longer sink short putts on the green easily).

People may have difficulty in swallowing if the sphincter at the bottom of their oesophagus (where it joins into the stomach) won't relax (achalasia). But a poison that 'causes fatal food poisoning' can help the food go down.

menace to medicine: pain

Regular botulinum toxin use has been moderately successful in the treatment of carpal tunnel syndrome, chronic backache, migraine, neck pain following whiplash injury, and other pain-related conditions.

But Keith Foster has modified the toxin to make it more effective against pain. He did this with his team from the Centre for Applied Microbiology and Research at Porton Down in the United Kingdom.

We have no idea why, but a protein from the Mediterranean coral tree, *Erythrina chritsagalli*, binds to the surface of nerves that carry pain, but no other cells. Foster's team married this protein to botulinum toxin. The combined chemical affects only pain nerves, and stops them from sending their signals, while ignoring all other nerve cells. Other nerves (that carry information to tell the muscles to move or receive sensation) continue to work normally.

Keith Foster's team found that his combination chemical was as effective as morphine at killing pain. But whereas morphine wears off after a few hours, his new chemical was still killing pain after nine days.

menace to medicine: incontinence

Urinary incontinence is the inability to properly control the passing of urine. And you guessed it, botulinum toxin can sometimes help.

Beatrice F. Brunger, 79, of Chicago had been suffering from urinary incontinence since 2000. The muscle walls of her bladder were over-active, which forced her to go to the toilet as often as four times a night. She could not get back to sleep, became permanently tired, and was getting weaker all the time.

The usual drugs for this type of incontinence did not work. The next stage would usually be an operation to make a 'holding tank' for the urine, via a hole in the side of her bladder.

But Dr Gregory T. Bales, a Urological Surgeon from the University of Chicago Hospital, used 'sausage poison' instead. He passed a small tube (with a camera and a tiny syringe inside) up through Mrs Brunger's urethra. Once it was inside her bladder, he injected the inside walls of her bladder with three vials of Botox (that's over three times the amount given to smooth out your forehead). He made 25 separate injections in about five minutes.

Mrs Brunger came home an hour later, and since then, has slept through the night. The only side effect she has encountered is that it takes her a little longer to urinate in the daytime, but she is happy with this. However, this application is so new that Dr Bales doesn't know how long the Botox will last, nor what its long-term side effects will be.

menace to medicine: cerebral palsy

Cerebral palsy is the most common disability in Australian children, occurring in roughly one in 400 births.

Although intellectual function may not be affected, the muscles are. A child's disability can range from a slight limp, up to being unable to walk or talk, or use their hands. The leg muscles can be in such a state of spasm that they become permanently shortened, so the child cannot place their heel flat on the ground. If the condition remains untreated, it can deform the leg bones. The botulinum toxin weakens the shortened muscles, so they can stretch. And of course once the child is able to walk, he gets a better sense of balance.

Originally, it was thought that a child being starved of oxygen during the birthing process caused cerebral palsy. We now think this occurs in fewer than 10% of cases.

A mother having a major infection during pregnancy can also cause cerebral palsy, as can the child having bleeding in the brain, low blood sugar levels, severe jaundice, etc. The bottom line is that in most cases, we don't know the cause of cerebral palsy — but in some cases, botulinum toxin can help.

physical side effects

All medications, including herbal ones, have side effects. The side effects of botulinum toxin tend to be local, depending on where they are injected. The side effects include difficulty in swallowing, headache, neck pain, drooping of the eyelids, nausea, and bruising or soreness of the injection site.

Surprisingly, botulinum injections could sometimes encourage wrinkles to form in untreated parts of the face. We don't know how this happens, but here's one theory. You take the botulinum injection, which temporarily paralyses some muscles. Your brain soon realises that a conditioned or programmed activity (e.g. frowning in annoyance, or lifting your forehead in surprise) is being blocked. This probably activates other muscle groups, which try to maintain the conditioned activity. So you lose wrinkles in one part of your face, only to grow new ones in an area that never had them.

When botulinum toxin is used to treat neck muscle spasm, the side effects included difficulty in swallowing (9–90%) and excessive muscle weakness (3–31%). In one case, the doctor shoved a needle too deeply into a patient's neck. This patient lost the ability to swallow, and had to spend three weeks in hospital being fed through a needle in her arm. In another case, a soap opera actress lost her voice for three months. The scriptwriters had to write her out of the series creatively until she recovered her voice.

Botulinum toxin is applied by many small injections, rather than a few big ones. So sometimes, one eyebrow could be up, while the other one is down, leaving you looking quizzical.

Pain is another side effect. I received an email from a woman who had hyperhydrosis (excessive sweating) on the hands and in the armpits. She said: *'I received 20 needles in each hand and 10 under each armpit — the pain was indescribable. The end result is that for me it decreased the sweating by about 70% for the hands and 95% for the underarms.'* So the treatment worked, but the pain from the injections was horrendous.

Around 4.3% of people with hyperhydrosis who receive injections to reduce sweating in the armpits get increased sweating in other areas. However, they were so pleased with the reduction in sweat from their armpits, they continued with the treatments.

Sometimes botulinum toxin can make you look worse. It can remove wrinkles — if they're caused by uptight muscles — but it will do absolutely nothing for wrinkles caused by the skin losing its elasticity. This happens naturally with ageing, smoking or sun exposure. In these cases, Botox can only do harm.

Pongy Pits Bad

Hyperhydrosis means 'sweating too much'. It can be caused by over-activity of the thoracic sympathetic ganglion chain (a chain of nerves running along the side of the spine).

It usually first appears after puberty, and sometimes goes away by itself in the sufferer's late twenties or early thirties. The sweat usually appears on the hands, feet and armpits. The sufferer may have to change their shirt four times a day, and may be unable to handle anything that must stay dry (such as paper). If you shake hands with them, you'll get wet hands. Even at rest, and at room temperature, they can sweat as if they've just run a marathon.

Botulinum toxin has had success in treating excessive sweating, usually in the armpits. Here it works directly on the nerves, not the muscles. In one study, it reduced sweating from 195 mg of sweat per minute, down to 24 mg per minute. Even six months after treatment, the sweating was still half the previous maximum.

frozen side effects

Another side effect of botulinum toxin is that it prevents a person from expressing their emotions physically. Botox 'freezes' your face. Your face can no longer express its full and subtle range of emotions, such as love and hate, or grudging approval or wry amusement. How can you tell what a person's reaction is when they have frozen smiles, unfurrowable brows and paralysed cheek muscles? Alex Kuczynski from *The New York Times* quoted film director Baz Luhrmann, who said that many female actors using this poison find that *'... their faces really can't move properly'*. Director Martin Scorsese agrees.

Paul de Freitas, a London-based casting director, says one-third of the actors he sees have had Botox injections. *'We waste a great deal of time weeding them out of auditions,'* he said.

In a *Sun-Herald* article, Jeremy Zimmerman, another London casting agent says that men are also using Botox, *'I had to veto Mickey Rourke for the leading role in a British film. I had to explain to his agent that we wouldn't be using him because his face looked so frozen.'*

Even so, Botox use is very common. Alex Kuczynski wrote in *The New York Times*, *'The television industry depends on Botox as much as it does on pancake makeup and forgiving lighting.'* Botox was the reason that *'a 47-year-old woman (Greta Van Susteren, an anchor for the Fox News Channel) had ... the countenance of a 25-year-old woman'*.

emotional side effects

Botox injections can also cause the 'Dorian Gray' effect. As botulinum toxin wears off, you have to constantly keep topping it up every three to six months. If you don't, the face returns to its original wrinkly state, with an additional three to six months of extra ageing. According to one doctor, *'You could marry a woman with a flawlessly even face, and end up with someone who, four months later, looks like a Shar-Pei (a dog with an incredibly wrinkled face, with many large rolls of fat).'*

What happens to a person's mind when they fall in love with another person, only to see them rapidly age over the next three months? What do you think and feel when your true love's skin changes from a baby's smooth skin, to the wrinkles of a bulldog?

Dr Peter Misra wrote in the *British Medical Journal*: '*23% of patients seeking treatment with botulinum toxin at a dermatology clinic had body dysmorphic disorder, and psychotherapy was considered the more appropriate treatment for them.*' In other words, a quarter of the people who wanted Botox didn't need it — they really needed psychotherapy.

Most men have always had difficulty in trying to understand women. 'Sausage poison' injections will make it even harder to read a woman's face, as her ability to express many of her subtle emotions is gradually destroyed. In some parts of American society, it's rare to see a woman over the age of 35 who can show anger in her face.

You Are Your Face

By the time people reach 50, they will be able to have the face they can afford.

And what's wrong with that? Coco Chanel, inventor of Chanel No. 5 perfume and posh apparel, once remarked how a person's character is etched on their face: '*At 20 you have the face that nature gave you, but at 50, you have the face that you deserve.*'

Perhaps your face should reflect what you have done, and what your experiences have turned you into.

Public Outing

Botox has been used by Renée Zellweger, Jamie Lee Curtis, Liz Hurley, Cliff Richard and the singer Lulu. Botox is a paralyser.

Collagen is quite different — it's a natural 'plumper-upper'. It's made from the connective tissue of pigs or cows. Collagen is injected into the skin to plump it up and to fill in the lines and creases. It's also used to enhance lips. Demi Moore and Barbara Hershey have used collagen injections. About 3% of the population are allergic to animal collagen, which can last from a few months to over a year.

Restylane is a synthetic, non-animal, collagen — it is stabilised hyaluronic acid. Like collagen, it's injected into the skin to soften and smooth, and to add volume to the lips. It lasts for about six months and has been used by Liz Hurley, Angelina Jolie and Calista Flockhart.

These three treatments are the fastest-growing in the injectibles market of the cosmetic surgery industry.

A new product currently being promoted is Isolagen. It's your own collagen cells, which have been grown in a laboratory and then injected back into your face. It's fairly new, but it's already thought to be longer lasting than Restylane or collagen.

The various faces of Botox
Look at me,
I'm crying!
I am despair!
This is angry!
No. 1...can you
cry for me...?
Casting List
No. 2...give me
extreme despair...
No. 3...anger
give me anger...!
LA screen tests — Botox style

long-term side effects

In the early 21st century, we don't know the long-term effects of injecting botulinum toxin into people. But we will know after a decade or two.

Back in the early 1500s, the famous doctor, Paracelsus said: *'All drugs are poisons, what matters is the dose.'* What are the long-term effects of getting this toxin every three months for 20 years? Remember the old medical saying: *'Never be the first, or the last, to try a new treatment.'*

On the other hand, botulinum toxin has worked for singer, Sir Cliff Richard. His perpetually young face is mostly due to the most powerful poisons known to the human race. Will botulinum toxin be the Elixir of Youth (albeit temporary), or will it be the Medical Botch-Up Of The Future?

References

'Actor on Botox? We'll call you ...', *The Sun-Herald* (Sydney), 16 February 2003, p. 37.

Blackie, Jeff, 'Botulinum toxin type A — therapy for neurological disorders', *Current Therapeutics*, March 1993, pp. 11–14.

'Botulinum toxin soothes chronic pain', *New Scientist*, 19 April 2003, p. 14.

Chin, James (Editor), *Control of Communicable Diseases Manual*, American Public Health Association, 2000, pp. 70–75.

Kuczynski, Alex, 'In quest for wrinkle-free future, frown becomes thing of past', *The New York Times*, 7 February 2002.

Misra, V. Peter, 'The changed image of botulinum toxin', *British Medical Journal*, vol. 325, 22 November 2002, p. 1188.

Naumann, M., Lowe, N. J., Kumar, C. R. & Hamm, H., 'Botulinum toxin type A is a safe and effective treatment for axillary hyperhydrosis over 16 months', *Archives of Dermatology*, vol. 139, June 2003, pp. 731–736.

Schittek, Birgit, *et al.*, 'Dermcidin: a novel human antibiotic peptide secreted by sweat glands', *Nature Immunology*, vol. 2, no. 12, December 2001, pp. 1133–1137.

Woods, Kate, 'Botox may lead to further wrinkles', *Medical Observer,* 7 March 2003, p. 41.

Thanks to George Papanicolaou and his wonderful wife, Andromache, millions of women today are being saved from cancer of the cervix. And it all started with George smearing vaginal cells from guinea pigs onto a glass slide ...

mrs pap smear

Today, the second most common form of cancer that kills women in Western society is cancer of the cervix (breast cancer is number one).

But thanks to the famous Pap Smear, we can detect cancer of the cervix before it digs in locally and/or spreads throughout the body, and successfully treat it. The Pap Smear is named after Dr Papanicolaou, who did a Pap Smear on his wife virtually every day for 20 years.

I once spoke to a woman who was diagnosed with cancer of the cervix when she had her very first pap smear at the age of 18. Twenty years later, she can confidently expect to live until 78, just like other Australian women. This woman owes her life to Andromache Mavroyeni, who married George Papanicolaou.

life of george

George Nikolas Papanicolaou graduated with a Medical Degree (with Honours) from the University of Athens in 1904. He then spent two years as an assistant surgeon in the Greek Army, followed by a year in a leper colony. He was more interested in medical science than medicine, so he left for Germany and enrolled as a PhD student in biology at the prestigious Zoological Institute in Munich. He again did well in his studies, and by 1910 had received his Doctorate and returned to Greece.

Then something wonderful happened to George, which enriched his life and saved the lives of tens of thousands of women. On a ferry boat journey to Athens, he met Andromache Mavroyeni. ('Andromache' means 'a woman fighting with men'.) Andromache was clever and well educated, could speak French, play the piano, and had a great personality. So they married on 25 September 1910 and headed for Paris for a holiday.

When his mother died, George returned home. A few months later, the Balkan War broke out and George re-entered the Greek Army where he met many Greeks who had returned from America to fight. They convinced him to emigrate to the USA, where he could make his dreams come true.

USA — the land of opportunity

George and Andromache landed in New York on 19 October 1913. By September 1914 George and Andromache found themselves working at Cornell University, in the Anatomy Department, which was headed by Dr Charles Stockard. The name 'Andromache' was a little difficult for Americans to spell and pronounce, so she took the name of Mary. 'Mary' and George worked together as scientific colleagues for the next 47 years.

Stockard's department was studying what effect long-term alcohol exposure had on guinea pigs and their babies. In special tanks, animals were born, bred and died in an atmosphere thick with alcohol fumes. George, who always held an interest in sex determination and differentiation, got permission to work with a few spare female guinea pigs. For his research, he needed guinea pig eggs just before ovulation (when the eggs are expelled from the ovary). But nobody knew how to tell when a guinea pig was ovulating or not. One morning he woke up with a Cosmic Flash, which he wrote down: *'The females of all higher animals have a periodic vaginal discharge: so lower animals such as these rodents, should also have one, but one probably too scanty to be evidenced externally.'*

the great discovery

All George had to do was look and take vaginal samples every single day to see what changed. So, on his way to work that morning, he dropped in at Tiemann's Surgical Supply shop. He bought a small nasal speculum to look at the vagina of a guinea pig. He collected vaginal cells, and spread (or smeared out) the vaginal debris and cells on a glass slide, then examined them under a microscope. (This is why it's called a Pap SMEAR.) Thanks to his knowledge of cells, George could see that these vaginal cells were at different stages through their life cycle.

George was incredibly excited by what he saw and wrote in his diary: *'There were moments of excitement when the examinations of the first slides revealed an impressive wealth of diverse cell forms and a sequence of distinctive cytologic patterns.'*

That evening, he somehow convinced his incredibly obliging (and brave and scientifically curious) wife, Andromache, to consent to the very first Papanicolaou Smear. He was delighted to see patterns in her vaginal cells that were similar to the patterns he had seen in the guinea pig vaginal cells that morning.

Luckily for science (and for millions of women), his wife consented to being a long-term volunteer. He did a Pap Smear on Andromache every day — and the days turned into weeks, months, and eventually years. Thankfully, Andromache had uncomplicated and regular cycles. He observed the normal changes in the cells in different parts of her vagina through one menstrual cycle; then from one menstrual cycle to the next.

the long path

But the Pap Smear really comes into its own when you find an abnormality in the cells. Dr Pap's work with his wife was only the beginning of what was to be a very long journey before his Pap Smear was accepted.

Back in 1917, George couldn't convince other doctors and scientists of the benefits of the Pap Smear. He soon realised the changes he saw in the cells over a menstrual cycle were directly related to changes in the ovaries, the uterus, and various female hormones. Other scientists read his paper in 1917 describing his work on the guinea pig, and as a direct result of his work, they soon discovered many female hormones. By 1919, George had built up an excellent reputation in the scientific community.

George was regularly using his smear method as a diagnostic tool in women. In 1923, he began to study vaginal changes in newborn babies and children, in fertile and menopausal women, and in women with hormonal abnormalities. In February 1925, while at the Women's Hospital of New York City, he recruited a dozen volunteers (mostly hospital staff) and began a long-term study of their vaginal cells. He collected cells by sucking them up through a glass pipette from the posterior vaginal fornix (at the back of the vagina, where the uterus enters the vagina). After a while, he realised he could get lots more useful clinical diagnostic information by examining the cells from the cervix.

Who Collected Mrs P's Pap Smear?

In the early days, George was rather discreet about his daily Pap Smear from his wife. In fact, he and Andromache kept quite a few details deliberately vague. Later on, George would introduce her as *'my wife and my victim'*.

In 1933, he wrote: *'Most of the smears which I examined up to February 1925, were obtained from a single human case (special case). This was an ideal case to work with, because of perfect regularity in the menstrual periodicity and of a complete lack of any serious disturbance or of any bacterial contamination. It has been by far the most normal and typical case that I have had an "opporunity" to study.'*

Was that little spelling error in the third-last word brought on by overwhelming love and admiration for the regularity and perfection of Andromache's Pap Smears? But Dr Claudia L. Barton from the College of Veterinary Medicine at Texas A&M made an amazing suggestion in her paper, *'The Historical Background of Cytology'*. She writes: *'... his wife Mary is reported to have collected vaginal smears from herself daily from the beginning of his work through her menopause.'* If she collected her own Pap Smears, she is truly a Goddess.

Pap Smear and Genetic Defects

Soon, Pap Smears should be able to indicate whether a 6-week old foetus of a pregnant woman has genetic abnormalities.

When a woman is pregnant, the placenta does a very good job of keeping apart the mother and the baby. But in 1971, it was discovered that 0.3% of the cells in the cervix of a pregnant woman came from the foetus. In 2003, Darryl Irwin from the University of Queensland worked out how to separate these foetal cells, and to screen them. This test for genetic abnormalities would be cheaper, quicker and less risky than other screening methods — and should be widely available by 2005.

a greater discovery

One day George examined a Pap Smear which had cells from a small cervical cancer. He later wrote: *'... the first observation of cancer cells in the smear of the uterine cervix was one of the most thrilling experiences of my scientific career.'*

Dr Papanicolaou's brilliant discovery was that it is possible to diagnose cancerous and pre-cancerous cells in the cervix, in a relatively painless way — and often at a very early stage before the cancer had spread. This information is very useful. Cervical cancer can be treated successfully, especially if it's diagnosed at an early stage.

George soon found other women with cells indicating they had a cancer of the cervix. Unfortunately, his scientific and medical colleagues still treated him with scepticism and indifference. But his wife, Andromache, continually encouraged him verbally (and scientifically) by letting him do a Pap Smear on her every day.

slow acceptance

In 1928, Dr Papanicolaou gave a lecture entitled *'New Cancer Diagnosis'* at the Third Race Betterment Conference in Battle Creek in Michigan. He described his technique and what he had found. But again, there was very little acceptance of his work.

In 1939, George teamed up with Herbert Traut, a gynaecological pathologist at Cornell. They took vaginal Pap Smears from all the women admitted to the Gynecology Department of New York Hospital, and George personally examined every smear. To the surprise of the rest of the scientific and medical community, but exactly as George had thought, he found a large number of previously-undetected early cervical cancers.

In August 1941, Papanicolaou and Traut published their famous paper, *'The Diagnostic Value of Vaginal Smears in Carcinoma of the Uterus'* in the *American Journal of Obstetrics and Gynecology.* This paper was the turning point for Pap Smear examination, and the medical profession began to take notice of this new technique.

In 1941, about 26,000 women died from cancer of the cervix in the USA. In the mid-1990s, the death rate had dropped to 4,800.

The reclining female
Spatula used for Pap Smear specimen collection
Speculum
Cervix
The slightly intrusive but highly informative Pap Smear procedure.
Thanks to the famous Pap Smear, we can find cancer of the cervix before it digs in locally and/or spreads through the body, and successfully treat it.
Dr George Nikolas Papanicolaou
(Dr Pap Smear)

Honours

George wrote some 150 scientific papers, and received many honorary degrees and awards. His portrait appeared on postage stamps of the USA, Greece and Cyprus, and on the 10,000 drachma Greek bank note.

Mrs Papanicolaou, who was the intellectual equal of her husband as well as a willing volunteer, was eventually recognised as late as 1969. The American Cancer Society gave her a special award and declared her 'A Companion In Greatness'. I personally reckon she deserved more.

cervical cancer — cause and growth

There are many different causes of cervical cancer, but most are associated with HPV — the Human Papilloma Virus. There are many different varieties of this virus. Types 1, 2 and 4 cause common warts on the hands and plantar warts on the feet. Types 3 and 10 cause flat warts. Type 11 causes warts on the larynx. Type 16 is associated with cancer of the cervix.

HPV Type 16 weakens the ability of the cells in the cervix to suppress or inhibit cancers. So once a cancer starts, it can grow rapidly, because HPV weakens the defence mechanisms.

Smoking, sexual intercourse and increasing age also increase the risk of cervical cancer.

The progress of cervical cancer follows a predictable sequence. On average, it takes at least a decade for a cervical cancer to develop fully.

The first indication is changes visible on the Pap Smear, but not to the naked eye. The progress to visible and invasive cancer is usually slow. Regular screening can catch cervical cancers in the early stages. If the screenings are frequent, cervical cancer is completely preventable, and totally curable.

Conventional Pap Smear

The classic conventional Pap Smear is simple and cheap.

The clinician removes some cells from the cervix and/or vagina with either a brush or a spatula. Next, they smear the cells thinly onto a glass slide — this is how it gets the name Pap Smear. Then the clinician preserves the cells by spraying them with an ethanol-based fixative. However, if some of the cells are in a layer only one-cell thick, they dry in the air before the fixative is sprayed on.

There are still debates about how best to actually smear the cervical cells onto the glass, and also about how to classify and manage the various stages of cancer of the cervix.

how often to screen

The risk of cervical cancer varies around the world. In Hong Kong, one out of every 72 women will get cervical cancer during her lifetime. The risk is fairly low until around 35, and reaches a peak at age 75.

Obviously, as you screen more frequently, you increase your chances of finding cervical cancers. However, testing an entire population is expensive, and luckily, most of the cervical cancers grow quite slowly.

The numbers vary with the population, but in Hong Kong they run like this. Screening every woman every year would probably reduce cervical cancers by 93.5% (as compared to not screening at all). Screening every two years would reduce it to 92.5%, while screening every three years would give you 90.8%. Screening every five years is much worse with a reduction of 83.6%, while screening every 10 years is significantly worse again at 64.1%.

Depending on the wealth of the country and the population, screening women every one to three years on average seems to be quite effective. Some women need more frequent screening.

wealthy countries vs poor countries

Even in wealthy countries today, where the Pap Smear is readily available, women are still being diagnosed with advanced cervical cancer.

Typically, about 50% of these women have never had a Pap Smear, while 10% did not have one in the preceding five years. Another 10% were poorly followed up, while another 30% were associated with reading errors of the Pap Smear, or sampling errors. This means 60% of cervical cancers in wealthy countries could have been avoided if women had been regularly screened.

But the Pap Smear is shamefully under-used in rural Aboriginal communities in Australia, and in poor countries. In poor countries, or where there is no access to Pap Smear screening, 5% of all women are affected by cervical cancer. This drops to about 1% where Pap Smear screening is available.

In Australia, Aboriginal and Torres Strait Islander women are not well screened for cervical cancer. As a result, cervical cancer is still the Number One cause of cancer death for these women. In fact, these women are 10 times more likely to die from cervical cancer than non-indigenous women. This is terrible, because cervical cancer is probably one of the most easily preventable cancers.

Stages of Cervical Cancer

Pathologists have graded cervical cancers into different stages, which indicate how far the cancer has progressed.

First is the **pre-cancerous stage**. This is when the Pap Smear sample can indicate pre-cancerous changes of the cervix. The survival is very close to 100% because it's not even a cancer yet. The treatment is to remove the pre-cancerous area by minor surgery, laser ablation, diathermy, etc.

Stage 0 Cervical Cancer is before it invades locally. At this stage, the cervical cancer cells live only in the epithelial cells right on the surface of the cervix.

Stage 1 Cervical Cancers have spread down into the connective tissue under the epithelium and are still confined to the cervix.

Stage 2 Cervical Cancers have actually left the cervix and spread elsewhere in the vagina, or to other tissues within the pelvis.

Stage 3 Cervical Cancers have spread to the lower part of the vagina or to the wall of the pelvis.

The most serious of all are **Stage 4** Cervical Cancers. They have spread to distant organs such as the bone, lungs or bladder.

Today, if a woman in a wealthy country has a cervical cancer that has not spread beyond the cervix (Stage 1 or lower), she has an 85% chance of living for five years.

the end

Dr Papanicolaou died in 1962, and his wife, Andromache, died in 1982.

Dr Charles Cameron said of Dr Papanicolaou: *'He was a giver of life: he is in the company of the great: he is one of the elective men of Earth who stand for all eternity like solitary towers along the way to human betterment. We are deeply in his debt.'*

We are especially deeply in debt to his wife, Andromache, who had a Pap Smear every day for 20 years.

In her case, once a day did not keep the doctor away.

References

Coory, Michael D., Fagan, Patricia S., Muller, Jennifer M. & Dunn, A.M., 'Participation in cervical cancer screening by women in rural and remote Aboriginal and Torres Strait Islander communities in Queensland', *Medical Journal of Australia*, vol. 177, 18 November, 2002, pp. 544–547.

Papanicolaou, George, 'The sexual cycle in the human female as revealed by vaginal smears', *The American Journal of Anatomy,* vol. 52, no. 3, Supplement, May 1933, pp. 519–637.

Papanicolaou, George & Traut, Herbert F., 'The diagnostic value of vaginal smears in carcinoma of the uterus', *American Journal of Obstetrics and Gynecology,* vol. 42, no. 2, August 1941, pp. 193–206.

Tirilomis, Theodor & Malliarou, Stella, 'Medical history and the European union: Papanicolaou and Asklepios', *Medical Journal of Australia,* 21 April, 2003, p. 880.

Vilos, George A., 'The history of the Papanicolaou smear and the odyssey of George and Andromache Papanicolaou', *American Journal of Obstetrics and Gynecology,* vol. 91, no. 3, March 1998, pp. 479–483.

Dentists love dental floss. Dentists have a saying, *'You don't have to floss all of your teeth, just the ones that you want to keep.'* But dental floss has a dark side — it can help criminals escape from gaol, and sometimes, it can kill you.

killer dental floss

Dental floss is a thread, usually made of nylon, but it also comes as a tape. You use dental floss to clean the crevices in between your teeth, where the bristles of your brush can't reach.

Dental floss has been around for a long time. Marks from dental floss have been found in the teeth of pre-historic humans and American Indians. Various peoples have used fine threads, such as silk, to clean between their teeth. But the modern re-inventor of dental floss was a New Orleans dentist called Levi Spear Parmly. Back in 1815, he recommended people should clean between their teeth with a thin thread of silk.

To floss, or not to floss . . .

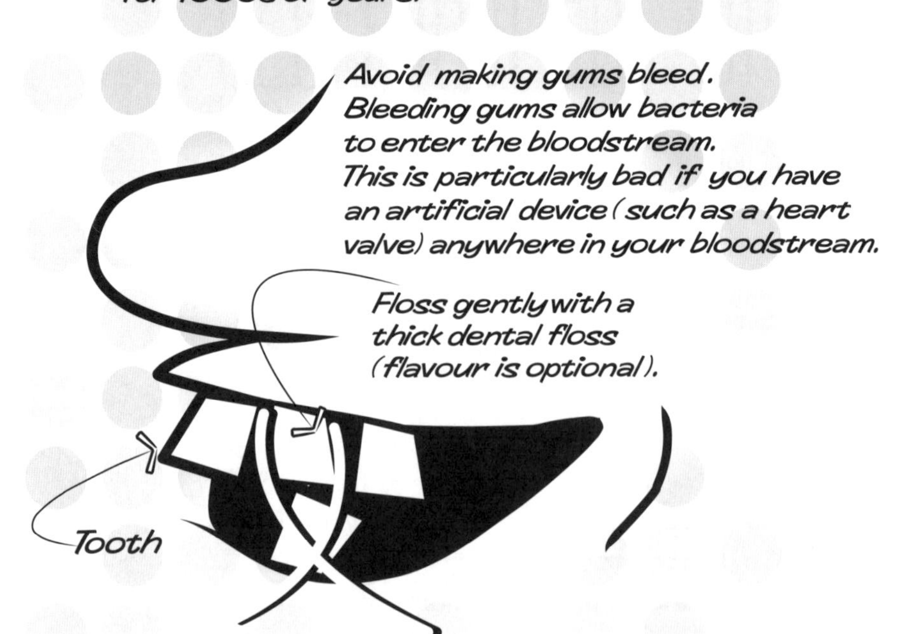

dental floss industry

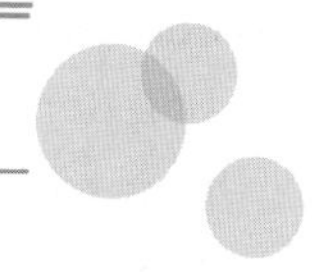

The Codman and Shurtleft Company of Randolph, Massachusetts, released the first commercially available dental floss in 1882. Their floss was an unwaxed silk. The Johnson & Johnson Company of New Jersey released their dental floss in 1896, and took out their first patent for it in 1898. Their original dental floss was made from the same silk as surgical sutures.

Wax-coated dental floss was introduced in the 1940s. During World War II, Dr Charles C. Bass, a medical doctor, realised that nylon had advantages over silk as a flossing material. Nylon could be produced in great lengths, and always with a consistent diameter. It could be supplied in different sizes, and was more resistant than silk to abrading and shredding. Dental floss as a tape, rather than a thread, was introduced in the 1950s, followed by flavoured flosses (such as mint and cinnamon). More recently, dental floss was made with various super synthetics such as Gortex and Teflon.

In 1996, Americans bought over 4.3 million kilometres of dental floss. Today you can get dental floss in all different types (from thread to tapes), in different diameters (with or without fluffy stuff), and with different amounts of wax and flavouring.

uses for dental floss

Dental floss has many non-dental uses. You can use it to slice cheese or remove biscuits that are stuck to the biscuit baking sheet. If you're really obsessive, you can clean the crevices in the turned legs of your wooden furniture. When you're on the road, floss can repair a backpack, tent or winter jacket. You can even tie together the legs of a bird for baking in the oven. Anaesthetists sometimes use dental floss to lock in place an endotracheal tube by using dental floss to tie the tube to a few teeth.

Of course the main use for dental floss is to clean those little nooks and crevices between the teeth that you simply can't reach with your toothbrush. You floss for the same reason you brush your teeth — to mechanically remove the build-up of bacteria from your teeth and gums.

Gum Disease and Genes

'Peri' means 'around' and 'dontal' means 'teeth', so periodontal disease happens around the teeth. And practically all of us have it.

The simplest type of periodontal disease is gingivitis, which means 'inflammation of the gums'. Gingivitis happens to 85–90% of the population.

Far more serious is chronic periodontitis, which affects 10–15% of people. The pathway runs from gingivitis to loss of the bone, and of the periodontal ligament (that holds the tooth in place). Ultimately, the tooth itself is often lost.

We really don't know why most people stay with the milder form of the disease, while a smaller group of people progress to the nastier form. But in some cases, your genes make you much more likely to get periodontal disease.

The gene is the IL-1 or Interleukin-1 Gene cluster. It makes Interleukin-1, which is a hormone that is active in inflammation but only over short distances. Interleukin-1 promotes the manufacture of enzymes that re-absorb connective tissue and bone. People who have an abnormal form of IL-1 suffer from gum disease.

But it gets worse. There's a link between periodontal disease and smoking. The chemicals in cigarette smoke somehow (we still don't know exactly) reduce your body's ability to repair itself from the ravages of periodontal disease. A 50-year-old smoker has the healing capacity of an 86-year-old non-smoker.

And it gets even worse again. There's a link between periodontal disease and heart disease. Again, we don't know why, but there are many theories.

The 'take-home message' is easy. Even if your genes are against you, you are still better off if you take care of your teeth (and that includes not smoking).

floss vs bacteria

One neat positive side to flossing is that in some cases it helps rid you of bad breath. The bacteria in your mouth break down any food particles they can find, and create volatile sulphur compounds. These sulphur compounds can give you bad breath. Remove the bacteria and sometimes you remove the bad breath.

But there's a more worthwhile reason for flossing.

If you don't remove the bacteria, they turn into 'plaque' on a tooth (a plaque is a film of mucous, in which bacteria live). After a while, this plaque turns into the hard substance called 'tartar'. Tartar provides a protective shelter for the more dangerous types of bacteria to grow. These bacteria make toxins, which irritate your gums and cause inflammation. This inflammation is called 'gingivitis'.

If you don't treat the gingivitis, you end up with 'periodontal disease'. Here the bacterial toxins invade and damage the gums, the bones and the structures that support your teeth. Periodontal disease is quite serious, and unfortunately, the symptoms often appear late.

floss vs caries

A final good reason for flossing your teeth is that it can help prevent 'caries'. Caries is a hole in the hard white outer enamel layer of the teeth. The enamel is the hardest material in the whole body. But it's very susceptible to acid.

'Caries' is a Latin word meaning 'rot' or 'rotten'. (The word 'caries' is a noun, while 'carious' is an adjective.) It's an odd word. In the English language, we do not say 'a rot' or 'a rotten', so we do not say 'a caries'. Instead, we talk about 'a rotten area' or 'an area of rot'. So the dentist might say that you have 'a carious area' or 'an area of caries'.

Dental caries is not a new disease, in fact it has been with us in all known societies. But in Europe and North America in the 1700s, a very different type of caries developed because peoples' diets changed. This was due to the development of the sugar cane industry in America, which led to the marketing of sugar as a food additive.

But it was only as recently as the 1970s that we began to really understand what happens when a tooth goes rotten, or becomes 'carious'. Today we know that in the early stages of dental caries, the tooth gradually loses its mineral content. Then bacteria invade the demineralised areas of the tooth.

There are five related factors involved in caries. Three of them make it worse. They are: first, a relatively high percentage of a bacterium called *Streptococcus mutans* living within the dental plaque; second, some kind of sugar in your diet; and third, frequent eating. The two protective factors are fluoride and lots of saliva.

Promote or Prevent Rotten Teeth

There are three factors that promote rotten teeth, and two that prevent them.

The first factor that encourages caries is a big colony of the bacteria, *Streptococcus mutans*, living inside the dental plaque. These bacteria make lots of lactic acid that attacks the tooth. They're also very well adapted for living and breeding inside that acid environment. If you have plaque with lots of *Streptococcus mutans* (2–10% by weight) you have a high risk of caries. But if there's less than 0.1%, you have a low risk of getting caries. Two other types of bacteria can also cause caries. They are *Actinomyces viscosus*, and several species of *Lactobacillus*. Like *Streptococcus mutans*, they make acid, and survive well in acid environments.

The second factor that promotes caries is lots of sugar in your diet. The bacteria involved in caries use the individual sugar rings, as well as the energy that holds the sugar rings together, to make chemicals called polysaccharides. These polysaccharides make the plaque thicker. They also change the liquid environment inside the plaque from liquid to gel. Before, when the plaque was more like a liquid, your saliva could diffuse it and dilute the acid. But now that it's more like a gel (jelly), the saliva can't penetrate, and the acid right up against your tooth's surface is less likely to get diluted.

The third factor that encourages caries is how frequently you eat — the more often you eat, the greater your risk of caries.

When bacteria come in contact with sugars in your mouth, they eat them for energy and make lactic acid as a by-product. The acid then dissolves into the teeth, causing demineralisation. But when your mouth is neither acid nor base (pH neutral), you have remineralisation, where the crystals in your teeth regrow using the phosphate, calcium and fluoride from your saliva. Normally you have a happy balance between demineralisation and remineralisation. But if you have thick gel-like plaque, and then eat sugar, within a few seconds the plaque turns acid and stays that way for two hours. The balance swings over to demineralisation. So the more frequently you eat, the more you're at risk of dental caries.

The first factor that protects you against caries is fluoride. Fluoride comes from drinking water, and from oral mouth rinses. Fluoride penetrates into the tooth, and changes its chemistry. Teeth impregnated with fluoride can be 10 times less soluble than regular teeth, so they resist the acid better, which makes it harder for demineralisation to occur.

The second factor that shields you from caries is saliva. The more you salivate the better. Your saliva can swing the balance between demineralisation and remineralisation over to the good side. So if you chew your food vigorously, you'll generate more saliva.

killer dental floss

Dental floss can keep your gums healthy by removing plaque from your teeth, but dental floss can also kill.

Before you can understand how dental floss can kill you, you need some background information.

Whenever you cut yourself, blood leaks out. But even though blood flows out, bacteria can get into your bloodstream, usually once the blood has stopped flowing. Normally, once bacteria get inside a blood vessel, they have nowhere to hide, even though you've got about 100,000 km of blood vessels. The invading bacteria are very rapidly hunted down and killed by your ever-vigilant immune system.

But if you have an artificial device (such as a heart valve) anywhere in your bloodstream, the flow around this valve is turbulent. It's not as smooth as it is in the rest of your blood system. There are tiny areas where the blood hardly flows. Once bacteria get into these stagnant areas, they are much less likely to be attacked by your immune system. So the bacteria can multiply in relative isolation, and can start up a colony.

bacterial endocarditis

Most of the time, the bacteria just stay at home. But sometimes, the bacteria will leave home. They will gush into the bloodstream in regular 'showers' (say, every 12 hours). If you look at the body temperature record of somebody in this situation, you will see very obvious 'spikes' in temperature. Their temperature will 'spike' as the bacteria flood into the bloodstream and 'visit' most organs around the body.

When bacteria set up a colony or two on the heart valves, this is called 'bacterial endocarditis'. There are over 1,000 cases on record of people getting 'bacterial endocarditis' after a dental procedure (usually a tooth removal), The risk of developing this bacteria is one in every 1,500–4,000 dental procedures. The dentist usually protects susceptible people with antibiotics.

A recent Australian case happened in 2001 to a 25-year-old woman with two artificial valves in her heart.

When she was a child, a bacterium had infected her and gave her rheumatic fever. This damaged her natural heart valves, which eventually needed surgery to install two artificial heart valves.

floss and bleeding gums

Three weeks before she turned up at her local hospital, the young woman started to use dental floss, but stopped after two weeks because it had made her gums bleed. She didn't know it, but the bleeding had let a bunch of bacteria into her bloodstream. They settled down on her heart valves and grew just nicely (thank you very much). She then came down with diarrhoea, rolling fevers and abdominal pain for about a week. Unfortunately, instead of getting better, she became more ill.

She then suffered new symptoms of more fevers, sore throat and a cough. She went to her local hospital, where the doctors immediately started her on antibiotics for the pneumonia she had now developed. However, the doctors were worried about the possibility of a bacterial endocarditis, so they transferred her to a larger hospital for close observation. You see, even though antibiotics will kill bacteria that are floating in the bloodstream, they don't work very well when the bacteria are living in a little colony. It seems that when bacteria get together, they make slimy chemicals that stop the antibiotics from invading the colony.

Sure enough, she gradually got worse and developed acute pulmonary oedema — liquid in her lungs. She was so sick she needed intubation and ventilation to keep her breathing, and cardiac support to keep her heart beating. She was then transferred to Alfred Hospital in Victoria.

The doctors in this teaching hospital found bacteria in her bloodstream. These bacteria normally live in the mouth. Obviously, when she had vigorously flossed her teeth, and caused bleeding, her mouth bacteria invaded her bloodstream.

The doctors also did an echocardiogram. It showed bacteria growing on her artificial mitral valves — both as small colonies and large vegetations. (The other artificial valve, the aortic, was untouched.) The surgeons immediately replaced the mitral valve.

The young woman recovered well, after being treated with industrial-grade antibiotics for several weeks.

flossing and heart valves

So if you're one of the few people who has an artificial heart valve, and you want to keep your teeth, you should floss frequently and gently, and use a large diameter flossing thread, rather than a skinny fibre.

Dentists have a saying, 'floss or die'. But for people with artificial valves, maybe the saying should be changed to 'floss right and live'.

Toothbrush — Electric or Manual

Electric toothbrushes are sold on the supposed advantage that they can clean your teeth better than you can with a manual toothbrush and a bit of elbow grease. The reality is that most electric toothbrushes are no better at cleaning your teeth than a manual toothbrush. On the other hand, electric toothbrushes are not worse than manual toothbrushes.

According to Professor Bill Shaw, of the University Dental Hospital in Manchester, an electric toothbrush is better than a manual one only if it uses one specific action. In his study, Professor Shaw looked at 29 clinical studies of the use of an electric toothbrush, involving 2,500 people. To do a better job, the brush head has to rotate clockwise and anti-clockwise quickly in short bursts. This action is called 'rotation-oscillation'. Only an electric toothbrush with this action does a better job of removing plaque than a manual one.

Rotation-oscillation toothbrushes remove about 7% more plaque, and lead to 17% less gum disease, as compared to manual toothbrushing. The other brush actions are:

1. side-to-side;
2. circular;
3. ultra-sonic vibration;
4. counter-oscillation, in which tufts of bristles spin in different directions at the same time.

None of these actions gave better results than a manual toothbrush.

The side effects of using an electric toothbrush are few, and there is very little evidence that it damages the gums.

Professor Shaw found that the single best way to avoid gum and tooth disease is to use a fluoride toothpaste. Over a three-year period, this reduces cavities in kids (aged 5–16) by 24%. He says, '... *whatever brush you use, use it properly, getting into the nooks and crannies and just under the gum line'.* He also said of electric toothbrushes, *'People are certainly fooling themselves, with the majority of [electric] brushes at least, if they think that they are more effective.'*

On the other hand, the newer electric toothbrushes with timers may encourage people to brush their teeth for a longer time — say, two minutes. I know that my electric toothbrush with a timer 'forces' me to brush for the full two minutes.

flossing and freedom

In some cases, dental floss has helped prisoners to escape from gaol.

A prisoner in a gaol near Palestine, Texas, coated his dental floss with toothpaste (which is mildly abrasive) and sawed through the steel bars in his cell.

Maybe he picked up the idea from the flight suits of combat fighter pilots. Depending on where they're fighting, their suits have different optional 'features'. One feature sometimes sewn into the suit is a diamond-coated thin steel wire (like piano wire) with two small rings at each end. If captured, they pick the wire out of their suit, loop it around a steel bar, put their fingers in the rings and quickly saw through the bar.

Another escapee was a gaoled Italian Mafioso called Vincenzo Curcio. He had committed one murder and arranged seven others, so the court threw him into the newly-built Vallette Prison in Turin. He discovered, and I'm not sure how, that the bars of this prison were made of a soft ductile iron. (Kids, make sure you pick up some metallurgy — a little knowledge can often help you in later life.)

Historically, most gaol bars have been made from an iron that's rich in carbon. This makes it very hard and difficult to saw through — however it does make the metal brittle. (It's fairly hard to persuade a metal to be both strong and flexible at the same time.)

In the 1970s, Italian officials were worried about terrorists breaking into prisons to release gaoled colleagues. They changed the specifications for the bars to be made of a softer iron, which would bend, rather than break, in an explosion.

Vincenzo Curcio went on a dental floss binge, presumably to keep his teeth clean. But secretly, when the officials weren't looking, he sawed through the bars of his cell using only his dental floss. And so he escaped, with probably the cleanest and healthiest teeth of any Italian prisoner.

That's one way to make a clean break for it ...

Strong vs Ductile

A material may be strong or ductile, but rarely both at the same time. In a metal, 'strong' means you need a lot of force to, for example, pull on it until it breaks apart. 'Ductile' (from the Latin, '*ducere*', to lead) means it can be drawn out or hammered thin, or easily fashioned into a new form. In other words, it's 'plastic' and will 'flow' if you apply enough pressure. (The opposite of 'ductile' is 'brittle', in which the metal shatters like glass.)

This is the problem with iron gaol bars — they can be either strong or ductile, but not both at the same time. However, in 2003, Drs Wang, Chen and Zhou described how they forced copper to be both strong and ductile at the same time. First, they rolled the copper between rollers at a temperature of around –204°C. Second, they heated the rolled metal to around 180°C. This treatment changed the previous single microstructure of the copper into two mixed microstructures.

About a quarter of the volume of this new copper is made of 'grains' about 1–3 microns (a micron is a millionth of a metre) in size. These grains can slide over each other, making the copper ductile.

The rest of the copper is made from much smaller 'grains' around 300 billionths of a metre in size (3–10 times smaller). These smaller grains won't split apart from each other, and so make the copper strong.

The treated copper was amazing. It had *'extraordinarily high ductility, but also retained its high strength'*. The large grains made it ductile, and the small grains made it strong. It could stretch by 65% before it broke.

The good news is scientists think that this technique can be applied to most metals — because their maths and computer simulations say so. This could mean a whole range of metals with improved properties may soon hit the market.

References

'Flossing To Freedom', *Fortean Times*, August 2000, p. 10.

Jenney, Adam W., *et al.*, 'Floss and (nearly) die: dental floss and endocarditis', *Medical Journal of Australia*, vol. 174, 15 January 2001, pp. 107–108.

'Murderer escapes by skin of teeth', *Sydney Morning Herald*, 15 April 2000, p. 31.

Valiev, Ruslan, 'Nanomaterial advantage', *Nature*, vol. 419, 31 October 2002, pp. 887, 889.

Wang, Yinmin, *et al.*, 'High tensile ductility in a nanostructured metal', *Nature*, Vol. 419, 31 October 2002, pp. 912–915.

It's true that birds of a feather flock together. You have probably seen a flock of birds flying in the sky, wheeling gracefully this way and that, and landing elegantly as a single unit, and wondered to yourself, *'How do they do that?'*

flocking birds

Does a flock of birds have a leader? And if they don't, how do they all decide democratically to do the same thing at the same time?

Try a little experiment on the flocking behaviour of pigeons. Gather a good supply of bread, some pigeons, and a friend to help you. Throw the bread to the pigeons every five seconds, but get your friend to do it every ten seconds. At the very start, half of the birds will come to you, and half to your friend. But within two minutes, the pigeons will realise that you are the better food supply, and two-thirds of the birds will cluster around you. When you and your friend swap your feeding rates, so that your friend gives them twice as much food, it will take the flock of birds another two minutes to readjust. Two-thirds of the birds will then cluster around your friend.

How do the birds know how to flock from you, to your friend, and back again — without a leader?

safety in numbers

We do know that all sorts of creatures flock — birds, bacteria, slime moulds, fish, whales, elephants and wildebeest, as well as sheep. It turns out that there are a whole bunch of advantages in being part of a flock.

First, you are less likely to be eaten by something bigger. Suppose that you are looking for fish out in the open sea. Your chance of finding a school of 10,000 fish is only slightly better than your chance of finding a single fish. Suppose that a predator discovers the school, and eats only one fish. Then the chance of any individual fish being eaten, is 10,000 times smaller when it's in the group, than when it's all on its own.

The 'confusion effect' of 'too many targets' also offers you protection. Duck shooters know that if you just fire blindly into a flock of ducks, your shot will almost certainly pass between the ducks. And hawks have been seen diving right into the middle of a dense flock of birds, missing every single bird, and coming out empty-clawed on the other side. (You have to aim at a specific target.)

There's another advantage if you are small, but there are a lot of you. You can sometimes band together collectively against a larger common enemy — and repel it successfully. In addition, the group/school/flock has more 'collective intelligence' to draw upon. People who spend a lot of time looking at European starlings returning to their roost in California have noticed that a small flock will get lost more frequently than a big flock.

If a few eyes are good, then more eyes are better. Bird scientists — ornithologists — have noticed that if a foraging bird that likes to hang out with its mates has to forage on its own, it will spend so much time looking over its shoulder for predators, that it won't get a decent feed.

Birds of a feather . . . flock where there is bread

Flocking Theory

It's really weird that animals can spontaneously organise themselves into what looks like very intelligent behaviour — and that it all happens as a result of following a small number of very simple rules. Even weirder, it turns out that you can get this seemingly-intelligent behaviour with quite different sets of rules.

Andreas Huth and Christian Wissel from Fachbereiche Physik und Biologie at Marburg University in Germany offer this set of rules to explain how a flock or school keeps together. These rules involve **repulsion**, **attraction** and **parallel orientation**.

The force of **repulsion** keeps the birds or fish a little distance away from each other. Normally the animals in a flock or school don't collide. If they do get too close to each other, they'll veer off to avoid collision.

But then the force of **attraction** comes in. If a creature is too far away from the school or flock, it'll head back in. The fancy name for this behaviour is 'biosocial mutual attraction'.

The third behaviour is **parallel orientation**. This just means that they all keep pointing in the same direction.

In their computer simulation, they let each fish interact with its four closest neighbours. When they ran these rules on their computer-simulated school of fish, they found that these three behaviours put together (repulsion, attraction and parallel orientation) kept the school of fish going in the same direction.

The school looked exactly like it had a leader that all the other fish followed — even though each fish in the school was just following these rules.

energy savers

Does flying in a flock save energy? Probably.

Flying is metabolically expensive. When a bird flies, it burns up a lot of energy — up to 12 times more than normal. There are claims that birds in a flock do get improved aerodynamic efficiency, because of the vortices, or spirals, of spinning air coming off the tips of the wings.

The best way to understand this is to look at aeroplane wings. Boeing 767s are notorious for having very powerful, and dangerous, wing-tip vortices. A moving wing generates lift because there's low pressure on the top of the wing, and high pressure underneath it. The higher pressure underneath pushes the wing upward, and the plane flies.

But the high pressure underneath the wing is not trapped there. At the tips of the wing, air spills off from the higher pressure zone underneath to the lower pressure zone on top. As the wing moves forward, it leaves behind a spinning whirlpool (or vortex) of air trailing off from each wing-tip. These separate vortices can sometimes converge behind the aeroplane and generate so much down-thrust, that a 'puddle-jumper' like a small Cessna flying several kilometres behind the Boeing 767 can be slammed into the ground.

If there's a down-thrust directly behind the aeroplane, then there are two up-thrusts — one on each side of the plane. The lift from that up-thrust means free energy. The same up-thrust occurs for birds flying in a flock. Supposedly, birds in a flock position themselves to use the energy in this up-thrust, which would otherwise be wasted. Some aerodynamic scientists have claimed that this up-thrust wouldn't really make much of a difference.

But consider the Alaskan bar-tailed godwit, which has the longest non-stop migration (11,000 km) of any bird. It eats until fat makes up 55% of its weight. Then, once it's fully loaded, the kidneys, liver and gut shrink enormously — which saves on wasted weight. For a migrating bird that goes these great lengths, perhaps even a small increase in flying efficiency is worth it.

'V' for '$'

In 2002, the NASA Dryden Flight Research Center at Edwards, California, proved that jet-fighters could save fuel by flying in a V-formation.

One NASA F/A-18 flew in the wing-tip vortices off to the side and behind another NASA F/A-18. At cruise altitude, the following fighter used 12% less fuel. This works out to either 600 lb less fuel used, or an extra range of 100 nautical miles.

are leaders really necessary?

An odd feature of most flocks or schools of creatures, is that they don't seem to have *one* single leader. If you look at a flock of birds wheeling in circles at sunset, you'll sometimes notice that the birds at the pointy end of the flock get left behind. Suddenly, the so-called 'leaders' have to do some quick flapping to return to the front of the flock.

It appears that flocks of birds run perfectly well without a leader, as do schools of fish and other creatures.

In fact, there are cases when organisations do just as well, perhaps even better, when the individual parts run themselves. For example, on 7 December 1991, the leaders of Russia, the Ukraine and Belarus issued a declaration: '*The Union of Soviet Socialist Republics, as a subject of international law and geopolitical reality, is ceasing its existence.*' That single announcement sounded the end of the Soviet Union, with its centralised power structure, after a run of nearly three-quarters of a century.

The very next day, on 8 December 1991, John Akers, the Chairman of the powerful IBM company, announced that IBM would no longer be run from the top. Instead, it would be split into many individual business units that would run themselves.

There are similarities between organisations and flocks of creatures, but how do you explain how a flock knows what to do? You could say that birds in a flock keep together just by maintaining the same speed and direction as the other birds around them, and also, by not smashing into them. But the physics of flocking behaviour is more complicated than that.

For example, consider the famous 'fountain' trick that a school of fish can use against a predator. Suppose a barracuda comes across a school of small tropical fish. As it heads directly towards the centre of the school, the small fish swim off to either side, creating a cavity around the barracuda. These fish have incredible acceleration — in a 50th of a second, they can reach a speed of 20 body lengths per second. So, like a fountain, the school of fish splits in two as the barracuda heads into it, and then recombines behind the barracuda. This flocking behaviour is not coordinated by a principal.

Why Birds Don't Collide

Another way to look at Flocking Birds is to think of Emergent Collective Behaviour — behaviour that emerges from a collection of creatures.

The birds have to follow only three separate rules:

1. Fly in the same average direction as your neighbours.
2. Stay close to the centre of the flock.
3. Don't get too close to any other bird.

As you can see, rule 3 is the most important to avoid colliding with other birds. It also helps that birds have really fast reflexes, and that their eyes are positioned on the sides of their heads. And if they do run into each other, rules 1 and 2 mean that they are travelling at roughly the same speed, so any injuries will be minor.

order vs chaos

Physicists are beginning to understand why birds fly in a neat flock, while mosquitoes buzz around in total chaos, thanks to research done by Naohiko Shimoyama, Yoshinori Hayakawa and their colleagues from the Research Institute of Electrical Communications in Sendai, Japan. They looked at various large and small creatures, which usually hang around in groups of 10–100 individuals. They measured these critters.

They then divided the body length by the velocity and the time between wing beats, and came up with a number they called G. For a mosquito, G was about 100 — and mosquitoes just wander around almost chaotically in their swarm. Sparrows had a G of about 1 — they sometimes wander chaotically, and sometimes gather in flocks. But cranes, which had a G of 0.1, fly in very ordered formations. Suddenly, with their simple number, G, the physicists had a way of predicting which creatures might flock in groups, and which creatures might be chaotic individualists.

the physics of flocking behaviour

However, it was John Toner from the IBM Research Center and Yuhai Tu from the Department of Physics at the University of Oregon, who used other theories of physics to devise a good computer model of flocking behaviour. Their scientific papers are full of complicated phrases such as *'Goldstone mode fluctuations'* and *'the projection of delta V bar perpendicular to X bar π'* — but it's easier to explain it all in simple English.

Their theory, based on three separate sub-theories, is able to model how a group of creatures can move together as a single unit, even if each creature can only see the other creatures immediately next to it.

First, they used the Physics of Bar Magnets. Bar magnets will line themselves up facing the same direction — birds in a flock will also line themselves up this way. Second, they applied the Physics of Dust Particles in a Fluid. Sometimes particles of dust in a fluid can move apart from each other — just like birds in a flock. The third theory of physics they applied was that of Convection of Heat — where a little packet of hot water can move through a boiling liquid. In the same way, birds can spread information about where they're heading, by gradually circulating through the flock.

When Drs Toner and Tu plugged in typical values of real birds into their computer model (using these three theories of physics), they came up with an incredibly realistic simulation of how birds move in a group — which proves the cliché that birds of a feather *do* flock together.

References

Cled, Jack, 'How do birds know how to fly in formation without colliding?', *Focus*, October 2000, p. 32.

Did You Know?, Reader's Digest (Australia) Pty Limited, 1991, pp. 100, 103–104, 123–125.

Partridge, Brian L., 'The structure and function of fish schools', *Scientific American*, vol. 246, no. 6, June 1982, pp. 90–99.

Toner, John & Tu, Yuhai, 'Long-range order in a two-dimensional dynamical XY model: how birds fly together', *Physical Review Letters*, vol. 75, no. 23, 4 December 1995, pp. 4326–4329.

We have two eyes on the front of our heads. Each eye gets a slightly different picture of the world, so this stereo image gives us the impression of depth. Our two ears pick up sounds at slightly different times, so we can work out where noise is coming from. And now we've discovered why ...

nostrils smell differently

This research was done by Noam Sobel and his colleagues at Stanford University in California and at the Ministry of Environmental Protection in Israel. They investigated at how different odours stimulate your smelling apparatus. Most of your 'smelling' is done with a part of your nose called the 'olfactory epithelium' — a patch of yellowish tissue high up in the roof of your nose. Normally, it does not have a very good airflow running over it, but this changes when you sniff deeply.

In this yellowish patch, there are sensory cells specially adapted for smelling, as well as supporting cells to hold the whole structure together. There are also gland cells that supply a wet layer of mucus to cover the yellow patch. Different chemicals in the air enter your nose, dissolve in this wet layer, and then excite the sensory cells.

Your human yellow patch of olfactory epithelium has about 40,000 sensory cells in each square millimetre, with a total area of around 250 mm^2 (about the size of your thumbnail) — a total of about 10 million cells that can detect odours. A rabbit, on the other hand, has around 120,000 sensory cells per square millimetre — a total of 100 million sensory cells — which, if placed side by side, have a total surface area greater than the skin of the rabbit's body!

the nasal cycle

It seems that you do not breathe equally through each nostril. This was first noted by Kayser, a German nose specialist, or rhinologist, in 1895 — and it's still somewhat controversial today. You see, we all have three sets of erectile tissue in each nostril — the same sort of erectile tissue found in the penis and the clitoris.

In many people, apparently the erectile tissue will swell in one nostril, while shrinking simultaneously in the other nostril. So while one nostril is passing a lot of air, the other one isn't. A little later the swelling switches to the other nostril. The complete cycle takes anything from 40 minutes to several hours to run. Only a small number of people have the classical 'one nostril with good airflow and one with bad airflow', with a regular rhythmical changeover. But most of us have some degree of rhythmical change of airflow from one nostril to the other. This is called the 'nasal cycle'. It seems that this nasal cycle gets weaker as you get older.

It also turns out that, if you lie on your side, then after about 12 minutes, the erectile tissue in the nostril on that side will begin to engorge and swell. This might be linked to sensors in your chest and pelvis.

Nose = Air-conditioner

To humidify the incoming air, the nose secretes about one litre of moisture a day.

This moisture is mainly in the form of a sticky mucus. This mucus acts like flypaper to trap tiny bacteria and particles. But if these bacteria were allowed to stay in place, they would rapidly lead to an overwhelming infection.

So the cells at the back of the nose are equipped with tiny cilia, which operate at about 10 strokes per second. These miniature hairs push the film of mucus to the back of the throat, where you swallow it. In the stomach, the powerful acids destroy most of the bacteria.

smelling in stereo

If you look at the odour chemicals that land on your olfactory epithelium, you can think of them as two types — those that dissolve quickly and those that dissolve slowly. The ones that dissolve slowly have their maximum effect in the nostril which has a slow movement of air. This gives the slow-dissolving chemicals time to absorb into the mucus covering the olfactory epithelium. But if the air is moving fast, they get whisked over the olfactory epithelium before they get a chance to dissolve.

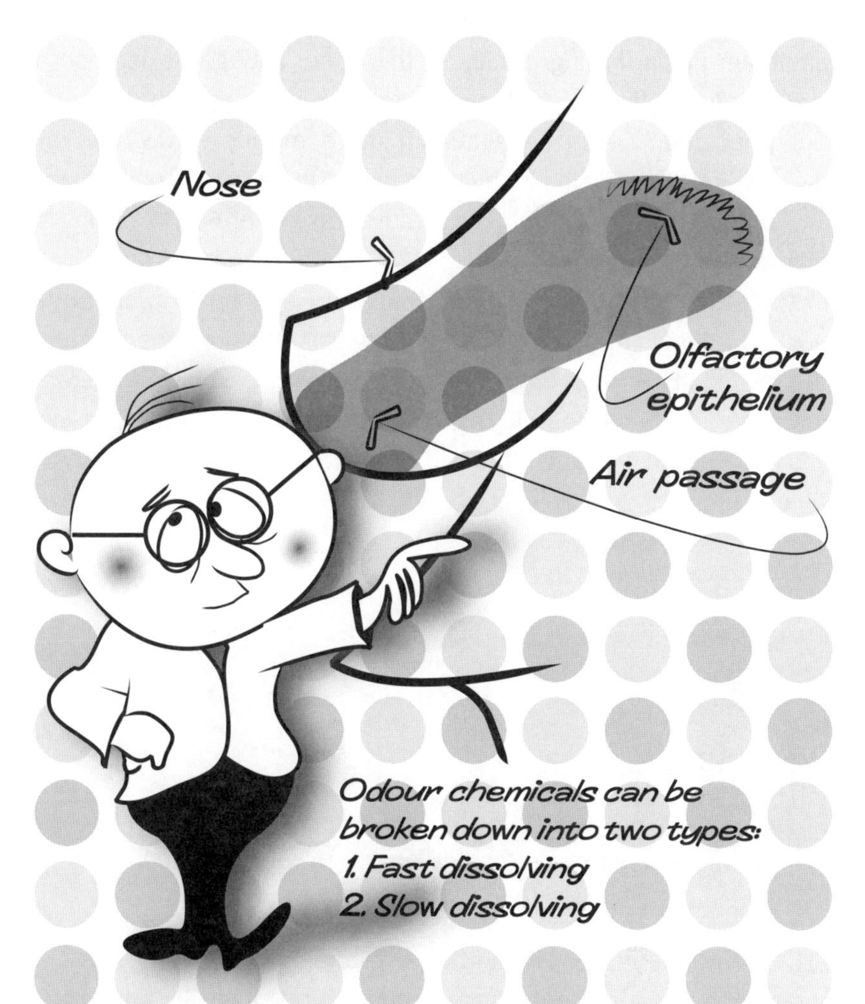

Chemicals that dissolve slowly have maximum effect in the nostril that has slow movement of air. Fast-dissolving chemicals have their strongest effect when they are in an airstream that is moving quickly.

Go on . . . take a sniff!

The story is completely reversed for the chemicals that dissolve quickly. They have their strongest effect when they are in an airstream that's moving quickly. In this case, they get to land on a large area of olfactory epithelium, and stimulate your brain more. But when they are in an airstream that's moving slowly, they all dissolve immediately in just a little patch of the olfactory epithelium.

We are not exactly sure what advantage there is in having a different airflow in each nostril. However, the researchers involved in the study offer this theory. If, thanks to the 'nasal cycle', one nostril is passing lots of air and the other is passing less air, then the same chemicals drawn into each nostril will give different responses in each nostril. This 'stereo' smelling ability will allow you to interpret the outside world better. It's food for thought next time something gets up your nose.

Baby Smell

Babies can smell before and after they're born — but not immediately after birth.

In the uterus, they float in amniotic fluid. Smell chemicals (such as from the mother's last curry) pervade this fluid, and trigger the baby's sense of smell.

Immediately after birth, the nostrils are clogged with amniotic fluid and other left-overs, and it takes a day for the baby to smell again. Like us, a baby likes the smell of honey and hates the smell of rotting eggs. In fact, a baby's sense of smell is better than ours. It can distinguish its mother's breast milk from another mother's breast milk.

wired nasally to the brain

There's another factor that makes your nostrils interpret identical smells in different ways — it has to do with how your nose is electrically wired to your brain.

Apart from stories about the swashbuckling character, Cyrano de Bergerac, the nose does not have a very romantic history. Very few poems are written to the nose, while thousands have been written to the eyes. All that happens to the nose is that you either pay through it, or you keep it to the grindstone.

Each day the nose cleans about 10,000 L of air, which has to be filtered and air-conditioned, so that it won't shock the delicate lungs. The air has to be modified to the climate of a hot humid summer day — about 80% humidity and a temperature of about 35°C.

Inside the nose there are coarse hairs that filter out most of the particles bigger than 10 microns — that's about one-seventh the diameter of a human hair. Particles smaller than 10 microns that make it past the hairs on the nose, land on the sticky mucus membranes inside the nose, and on the tonsils and adenoids.

Besides conditioning and cleaning the incoming air, the nose also processes smells.

Recent research into how the nose is wired to the brain was done by Larry Cahill and his colleagues from the Monell Chemical Center in Philadelphia, and the Center for the Neurobiology of Learning and Memory and the Department of Psychobiology at the University of California at Irvine. They demonstrated that identical smells coming into each nostril are treated differently in the brain, because of how the smelling areas in each nostril are connected to your brain.

left–right brain odours

The scientists exposed 28 male and female volunteers to eight different, but mildly pleasant odours — pineapple, coconut, maple, vanilla, peppermint, almond, lemon and anise. The volunteers were given these odours at two separate sessions, one week apart. The first time around, they smelt each odour through only one nostril, and at the second session, they smelt each odour through the other nostril.

When you sense a smell, the odour chemicals float into your nose, and land on the olfactory epithelium in each nostril. The chemicals stimulate the olfactory epithelium, which then sends electrical signals into the brain. If you look at the electrical wiring, you will notice that the electrical impulses from the left nostril travel to the left side of your brain, while those from the right nostril travel to the right side of the brain. This is a slightly 'fuzzy' statement to make, but overall, your left brain tends to deal with language and words, while your right brain tends to deal with emotions. The scientists thought that this left–right wiring might influence how your brain processes identical odours presented to each of your nostrils.

They were correct. The scientists found that each side of the brain did process the information from each nostril differently. When odours came in through the right nostril, the volunteers thought that they were smelling something more pleasant than when the same odour came through the left nostril. This agrees with our understanding that the right brain is involved in emotional processing.

But what does the left brain, which is involved in language, do to odours?

The scientists checked for this aspect by asking the volunteers to give a name to each odour that they were smelling. Sure enough, when the odours came in through the left nostril, the volunteers were more accurate linguistically and could give the correct name for the odour that they had been exposed to. So your brain tells you that odours are more pleasant if you sniff them through the right nostril, and your brain can name odours more accurately when you sniff them through the left nostril.

I wonder what's going on in your brain, when you think that you smell a rat?

Cold Nose, Warm Nose

We get a runny nose on a cold day. This is because the cold weather partially paralyses the cilia, and causes an over-production of the mucus. So the mucus can't be swept backwards to the throat, but instead dribbles out at the front.

The nose also warms up incoming air by running the air past three hanging bulbs of erectile tissue.

These erectile tissues can swell, and so provide a greater surface area for the incoming air to rub up against. This transfer of heat is so efficient, that if you were in the Snowy Mountains breathing in freezing air through your nose, the incoming air would be warmed up to the body temperature of 37°C by the time the air hits the back of the throat.

References

Bojsen-Møller, F. & Fahrenkrug, J., 'Nasal swell-bodies and cyclic changes in the air passage of the rat and rabbit nose', *Journal of Anatomy*, vol. 110, no. 1, 1971, pp. 25–37.

Herz, Rachel S., McCall, Catherine & Cahill, Larry, 'Hemispheric lateralization in the processing of odor pleasantness versus odor names', *Chemical Senses*, vol. 24, no. 6, December 1999, pp. 691–695.

Milius, S., 'Each nostril smells the world differently', *Science News*, vol. 156, 6 November 1999, p. 293.

Nelson Gilbert, Avery & Rosenwasser, Alan M., 'Biological rhythmicity of nasal airway patency: a re-examination of the "nasal cycle"', *Acta oto-laryngologica*, vol. 104, 1987, pp 180–186.

Sobel, Naom, *et al.*, 'The world smells different to each nostril', *Nature*, vol. 402, 4 November 1999, p. 35.

There are many different types of kisses — the soft fluttery kisses between a child and its parent, the chaste closed-mouth kiss given to a grandparent or an aunt, and of course, the wild, rollicking, hungry open-mouthed kisses between lovers.

right way to kiss

On average, we spend two weeks of our lives kissing. If we do it so often, how come one-third of us kiss the wrong way? Well, Onur Güntürkün got out there and 'observed' why.

We humans have a long history of kissing. Early Christians kissed whenever they greeted each other. The bride and groom kiss after the marriage ceremony. Eskimos and Polynesians kiss by rubbing noses together, while the inhabitants of south-eastern India kiss by each pressing the nose against the cheek, with the active person inhaling deeply. There's deep (or French) kissing where one person's tongue goes a'roaming in the other person's mouth — and we'll stop right there. And as a complete contrast, kissing in public was rare in Asia in bygone days, because the bow was the all-purpose greeting. People only kissed in private, so you wouldn't see it, of course!

Give way to the right when smooching

where did kissing come from?

There are a few theories about how kissing got so popular. One theory claims that it all began in ancient times when mothers chewed up food to pass it directly into the mouths of their babies.

A second theory reckons that kissing allows you to get close enough to smell the mood, the food and the recent adventures of the person you are kissing. This would help you work out how to handle them. After all, cats and dogs check out other cats and dogs this way. And what the heck, while we were up close and personal, we could touch foreheads, or noses, or perhaps even lips. So even after we no longer needed to sniff them out, we kept doing it because it was fun to do the other touching stuff.

And a third theory involves the belief that your soul lived in your breath, and that kissing would merge the breaths together, fusing your souls for all eternity. This last theory is really cute, but like all the other theories, it can't be proved.

In 16th-century England, a kissing game was very popular at the Elizabethan country fairs. A woman would press as many cloves as possible into an apple (the cloves acted as a breath-freshener). She would carry it around the fair until she found a kissable young man. She would offer him the apple, he would take and chew a clove, they would kiss, and he would then take the clove-studded apple on a mission to find a different kissable young woman. The kissing game would stop once the apple ran out of cloves — and by that time, a few hundred people have been kissed twice.

On the other hand, some people thought it wrong to kiss. The state of Indiana in the USA has a law making it illegal for a man with a moustache to *'habitually kiss human beings'*. In Hartford, Connecticut, it is still illegal for a man to kiss his wife on a Sunday. And in 16th-century Naples, in Italy, kissing was an offence that carried the death penalty.

So why should you kiss? Well, first, kissing can burn up 26 calories a minute, so it can be good exercise. Second, some (quite weakish) research shows that people who kiss their spouses goodbye in the morning, will earn more money than non-morning kissers. And third, it's nice, and it's free!

Seven Steps to Kissing the Right Way

1. Your breath should be clean. Brush your teeth and tongue beforehand. A mint might help, but get rid of it before you kiss.
2. Your lips should be slightly moist (not completely dry, and not drenched in drool). Avoid gloss which is too gooey, and remember that too much lipstick goes everywhere.
3. Being close is essential. Even if you intend to shut your eyes for the Big Kiss, keep them open until you have worked out which way to tilt your head.
4. Should you keep your eyes open or shut? This is a personal preference. If they're open, you can see what's happening to your partner's face (happy, really happy, or really really happy) and can change what you're doing to make them even happier. But if your eyes are shut, you can better 'drown' in the sensations.
5. Breathe through your nose! If you hold your breath, you will pass out, and if you breathe through your mouth, you will hyper-inflate your partner's lungs.
6. What about your partner's tongue? It's best just to follow the sensation of what feels nice. Lick the tongue, caress

the tongue, wrestle the tongue to the floor of the mouth, and maybe suck on the tongue. (But never ever bite.)

7. The open-mouth kiss is very intimate (some prostitutes will do 'everything' except kiss), so pay attention to the 'whole person'. Use your hands to make the other person happy. Cuddle, and always be comfortable.

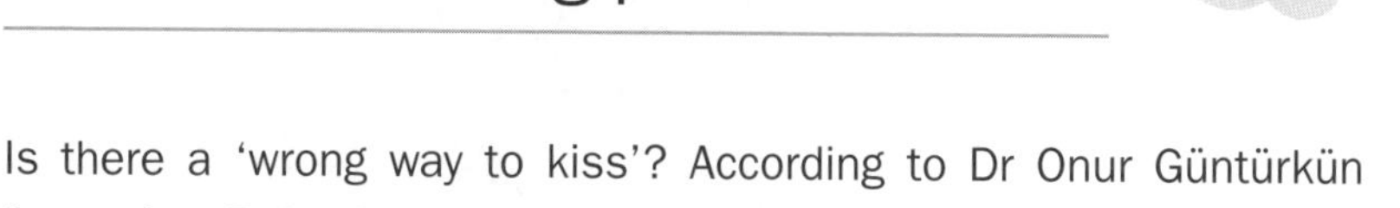

kissing pairs

Is there a 'wrong way to kiss'? According to Dr Onur Güntürkün from the Ruhr University in Bochum, Germany, there is. He specialises in studying differences between the left and the right sides of the body and brain in various birds and animals.

One day, while he was stuck at Chicago Airport for five hours, he observed a particular couple kissing each other. He noticed that they each tilted their head to the right, so that each person's nose was to the right of the other person's nose. He suddenly realised that he had just seen a left–right kissing preference, and that public places would be ideal for collecting data about this.

Being a good scientist (as opposed to a perv in a brown raincoat) he immediately established some criteria: there had to be lip-to-lip contact; the faces had to be aimed at each other; there had to be an obvious direction of head-tilting; and finally, they couldn't carry any luggage because this might influence the direction in which they tilted their head.

This made it difficult to collect lots of data. He says, *'Couples kiss much more rarely than expected. In Chicago O'Hare, I once waited five hours for two data points.'* Over the next two and a half years, he collected data on 124 scientifically-valid 'kissing pairs' at airports, parks, beaches and railway stations. The results were clear-cut — two-thirds of people tilt their heads to the right. This, of course, means that one-third of us kiss the wrong way.

Best cheek forward

what's going on?

First, about 90% of us are right-handed. About two-thirds of us (around 60–70%) prefer to use our right foot, right eye or right ear. So on average, we humans tend to use our right side. Second, we humans tend to turn our heads to the right (rather than to the left) for our last weeks in the uterus, and in our first six months after being born. Therefore Dr Güntürkün believes that humans start off with a preference for looking to the right, and pay more attention to events on our right. He also claims that this preference carries right on into our fertile years, so that when we kiss, we will tilt our heads to the right.

Perhaps for once this is too much science, and we would be better off listening to the words of the 18th-century Scottish poet, Robert Burns:

'Honeyed seal of soft affections,
Tenderest pledge of future bliss,
Dearest tie of young connections,
Love's first snowdrop, virgin kiss',

and go right ahead and kiss in whatever way seems natural.

Best Cheek Forward

In general, if we want to express emotion, we show the left side of our face — and if we want to show our power, or avoid showing emotion, we present the right side of our face to the world. This is usually the case, in both paintings and photos.

One study found that 68% of female portraits and 56% of male portraits showed the left side of the face. Another study found that people in family portraits would tend to show the left side of the face when asked to express as much emotion as possible. But when they were asked to be as unemotional as possible, they would swivel their bodies and necks to show the right side of the face. Professional models follow this pattern as well.

Does showing the left side of the face to show emotion influence the way we tilt the head to the right when we kiss? Maybe, maybe not.

References

Güntürkün, Onur, 'Adult persistence of head-turning asymmetry', *Nature*, vol. 421, 13 February 2003, p. 711.

'Kissing right', *New Scientist*, no. 2382, 15 February 2003, p. 20.

Konishi, Yukuo, *et al.*, 'Asymmetrical head-turning of pre-term infants: some effects on later postural and functional literalities', *Developmental Medicine and Child Neurology*, vol. 28, 1986, pp. 450–457.

Nicholls, Michael E.R., *et al.*, 'Laterality of expression in portraiture: putting your best cheek forward', *Proceedings of the Royal Society (London B)*, vol. 266, 1999, pp. 1517–1522.

'Self-recognition and the right hemisphere', *Nature*, vol. 409, 18 January 2001, p. 305.

Skiba, Martina, Diekamp, Bettina & Güntürkün, Onur, 'Embryonic light stimulation induces different asymmetries in visuoperceptual and visumotor pathways of pigeons', *Behavioural Brain Research*, vol. 134, 2002, pp. 149–156.

Since the 1960s, scientists have believed that the tiny Y-chromosome was relentlessly headed for oblivion, and that it would be gone in a few million years. But the latest research shows that, for all its faults, the Y-chromosome (and its nasty by-product, boys) is probably here to stay.

y-chromosome: waste- or wonder-land?

If you have followed the news over the past few years, you will know that we humans have just about mapped our own DNA. The big difference between boy and girl DNA is that one of the chromosomes is different — boys have a Y-chromosome, while girls don't.

Just about every cell in your body has the entire DNA needed to make another you. Our DNA is many things at once — it's a blueprint that can make a human being, it's a history book of our ancestry, it's a giant medical book, and it's a whole lot more. But in this story, let's just think of the DNA as being a blueprint.

Ape and Human

It turns out that, as far as DNA is concerned, we are only 1.2% different from chimpanzees. So out of our three billion letters in our genetic code, only 36 million are different between humans and chimpanzees.

We have learnt this by comparing DNA sequences between humans and apes. And we have worked out when we separated from them, in terms of evolution.

There was a single main species of two-legged primates about 15 million years ago (Myr). About 12–15 Myr the orang-utans split off from our common stock to evolve into a separate species. The gorillas then split off about 5–8 Myr.

We humans then left the main branch of primates about 4–8 Myr, leaving the bonobo and the chimpanzee on the remaining branch. They evolved apart from each other a few million years ago.

chromosomes are us

When a cell gets ready to split into two, the DNA also prepares to split itself into two. (This happens so that each new cell has its own DNA). The DNA, which is 2–3 m long, bunches itself up into 46 little packages. We call them 'chromosomes' because as early as 1848, geneticists would colour them with various dyes, to show up various features — and 'chromosome' means 'coloured body'.

If you look at these 46 chromosomes under a microscope, you will see that 45 of them are the same in boys and in girls, but that the 46th chromosome is different. It's quite large in girls where it carries about 1,500 genes, but in boys it's tiny, and carries only 78 genes. So guys, the next time you start thinking that you are the Lord of Creation, just remember that your throbbing manhood is made possible by just 78 genes. In fact, you can think of women as the luxury model with all the options, while guys are the cheap economy model without the air-con, power-steering and cruise control.

Y-Chromosome and Fertility

Over the past 300 million years, for some unknown reason, the Y-chromosome has degenerated. It has shrunk from about 1,500 genes to only 78 genes. Many of these genes are involved in sperm production. There are many copies of these genes, probably to help keep sperm production going if some of these copies are damaged (usually by random

Chromosome sales — direct to the public

mutation). But sometimes, the damage is too great. It seems that about 10–15% of male infertility problems are caused by damage to the sperm production genes on the Y-chromosome.

x-chromosomes good

Women have an advantage with chromosomes 45 and 46 because they are big, bold, matching X-chromosomes.

You will understand this advantage if you compare it to building a house from an architect's plans. Suppose that one set of plans has been made hard to read by smudges, burger stains and spilt coffee. Completing your building with these damaged plans would be a difficult task. It would be a lot easier with a spare set of undamaged plans. Chromosomes 45 and 46 provide a spare set of plans for each other. The fertilised egg uses these plans as a blueprint as it grows itself into a human girl baby over the course of nine months. If one X-chromosome has a damaged section, the growing embryo in the uterus can use an undamaged section from the other X-chromosome as a blueprint.

y-chromosome bad

But boys supposedly couldn't do this. After all, while their chromosome 45 is a big boofy X-chromosome, their chromosome 46 is a minuscule Y-chromosome that doesn't have a matching partner. If the Y-chromosome is damaged, it is unable to read the X-chromosome next door to find an undamaged section. So the Y-chromosome looks like a disaster waiting to happen. And indeed, boys suffer quite a few diseases (such as red-green colour blindness) which girls very rarely get.

y-chromosome badder

The news gets worse for the pathetic little Y-chromosome. It mutates very rapidly, thousands of times faster than other chromosomes. It seems that it was virtually identical to the X-chromosome about 300 million ago. In fact, there are a few short sections on the very ends of the Y-chromosome that are identical to the very ends of the X-chromosome — but over 95% of the Y-chromosome is very different from the X-chromosome. It has mutated massively. This 95% never swaps with its partner X-chromosome. It seemed that this 95% could never repair itself, and that any defects would appear in its human male owner.

Over the past 300 million years, the once-proud Y-chromosome has shrivelled from about 1,500 genes to its current miserly 78 genes. And scientists believed that if it kept shrinking and mutating at the present rate, it would be totally useless in just a few million years.

Yup, for a while there, scientists were convinced that the Y-chromosome was a genetic waste-land and junkyard. They thought that it was highly repetitive and mostly non-functional.

y-chromosome fully wicked

In June 2003, *Nature* published some reassuring research by Dr Page of the Massachusetts Institute of Technology, which changed the geneticist's attitude that the Y-chromosome was a Waste-Land to believing that it is a Wonder-Land. Dr Page and his team found that the Y-chromosome has a special trick — it doesn't need a matching partner in order to repair any damaged sections (or bits).

They discovered that large sections of the Y-chromosome are 'palindromes'. You might remember from your high-school English days that a 'palindrome' is a word or phrase, that reads the same in either direction. The word *'radar'* is a palindrome, as is the word *'level'* and the phrase *'Able was I, ere I saw Elba'*.

It is very exciting to know that sections of the Y-chromosome are palindromes. It means that if one section of the Y-chromosome is damaged, it may be able to find an undamaged version of the same section somewhere else in the Y-chromosome — and use it to make some part of a human.

While this is good news, it is also a little disturbing. Yes, the Y-chromosome can fix itself up — but only by having sex with itself.

Y-Chromosome — Hall of Mirrors

There are about three billion 'letters' in the entire human DNA, spread over 46 chromosomes. About 50 million of them live on the Y-chromosome. About six million of these are palindromes. They are mirror-image copies, and are a kind of backup architect's plan. They can be used as a spare blueprint, if the other DNA letters in the Y-chromosome are corrupted. The longest single sequence is about three million DNA letters long.

References

Pääbo, Svante, 'The mosaic that is our genome', *Nature*, vol. 421, 23 January 2003, pp. 409–412.

Pagán Westphal, Sylvia, 'Decoding the Ys and wherefore of males', *New Scientist*, no. 2400, 21 June 2003, p. 15.

Rosen, Steve, *et al.*, 'Abundant gene conversion between arms of palindromes in human and ape Y chromosomes', *Nature*, vol. 423, 19 June 2003, pp. 873–876.

Skaletsky, Helen, *et al.*, 'The male-specific region of the human Y chromosome is a mosaic of discrete sequence classes', *Nature*, vol. 423, 19 June 2003, pp. 825–838.

Willard, Huntington F., 'Tales of the Y chromosome', *Nature*, vol. 423, 19 June 2003, pp. 810–813.

The very first use of the word 'software' occurred in 1850 in the rubbish trade — and it seems at times that it probably should have stayed there.

software sucks

Computer software is everywhere in our modern society. It runs sewage systems and banks, traffic lights and hospitals, electrical power grids and your car's engine, as well as its airbags, and probably even its stop lights. And when the software fails, it can make a real mess.

In 1996, faulty software made the French Ariane-5 rocket explode just 40 seconds after lift-off on its maiden voyage, destroying both itself and its billion-dollar satellite payload. Faulty software in computer-controlled radiotherapy machines fed massively lethal overdoses of radiation to cancer patients in the USA and Canada between 1985 and 1987, and again in Panama in 2000 and 2001. And in 1999, both the Mars Polar Lander and the Mars Climate Orbiter were destroyed by computer bugs. Computer bugs in badly-written software also delayed the launch of the very expensive 88-km^2 Denver airport for a whole year, and also crashed ambulance control systems in London, contributing to the deaths of up to 30 people.

software types

Software is the set of instructions that tells a computer what to do. There are two main types of software. First, there is 'system' software which controls the internal functioning of a computer via its operating system (such as a Windows operating system like ME or XP, or Apple's Operating System X, or the old and very reliable UNIX operating system). System software also controls essential peripheral items such as storage devices (like CD-ROMs and hard drives), monitors and printers. The second type of software is 'application' software, which basically processes data for the user. Application software can manipulate pictures, turn the computer into an intelligent typewriter, write spreadsheets, control databases, play games, edit movies, play music and even manage a payroll. Actually, there is a third category of software, called 'network' software, which helps computers linked in a network speak to each other.

so why is it called 'software'?

The very first use of the word 'software' occurred in 1850 — used in the rubbish trade. According to the *Oxford English Dictionary: 'Two other departments* [among rubbish-tip pickers] *called the "soft-ware" and the "hard-ware" are very important. The former includes all vegetable and animal matters — everything that will decompose.*' But it was John W. Tukey, Professor of Statistics at Princeton University who first used the word 'software' in its modern sense when he wrote an article in the January 1958 issue of *American Mathematical Monthly*. He wrote: *'Today, the "software" comprising the carefully planned interpretative routines, compilers, and other aspects of automative programming are at least as important to the modern electronic calculator as its "hardware" of tubes, transistors, wires, tapes and the like.'*

computer bugs

Anyone who has ever used a computer knows that one of the big problems with software is that it contains bugs. In computer terms, a 'bug' is some sort of fault in either the electronics or in the program (usually the program) that makes it do something bad. Watts S. Humphrey, a fellow of the Software Engineering Institute at Carnegie Mellon University, has spent many years analysing some 13,000 programs. He believes that good software should be *'usable, reliable, defect free, cost effective and maintainable. And software today is none of those things. You can't take something out of the box and know it's going to work.'*

He says that a professional coder or programmer will make about 100–150 errors per 1,000 lines of code of program written — and after error-checking, there remain about 0.5–5 errors per 1,000 lines of finished code. So the very popular Windows NT operating system, which has 16 million lines of code, probably has between 8,000 and 80,000 mistakes or bugs.

'a bug that fear'd us all?'

Today, we all rely on computer software, even if we don't have a computer. For example, when you make a phone call, you use about two million lines of computer code which took 1,000 person-years of effort to write. A single error, or bug, in any one of these two million lines could terminate your phone call. This software probably doesn't have any bugs in it. But each year, according to the Research Triangle Institute of North Carolina, bugs in software sold to industry cost the US economy about US$60 billion.

The word 'bug', with the meaning of 'something wrong or undesirable', has been around for a long time. In Shakespeare's *Henry VI*, King Edward refers to Warwick as '*... a bug that fear'd us all'.* Thomas Edison, who helped introduce the phonograph, the electric light bulb and the motion picture industry, used the word 'bug' to mean '*any fault or trouble in the connections or working of electric apparatus*'.

A 'Bit' Goes a Long Way

Professor Tukey from Princeton gave us more than the word 'software'.

He also gave us the word 'bit' — the smallest piece of information that a computer can handle. The word comes from 'binary digit'. In computers, it is usually either a '1' or a '0'.

Professor Tukey also played a big part in Fast Fourier Transforms, which can find patterns in data. And he was also involved in developing the statistical technique called Analysis of Variance, which can tease simple information out of complicated data.

is a moth a 'bug'?

But 'bug' really became popular because of Grace Brewster Murray Hopper, one of the greats of the computer industry. She was born in New York City in 1906, and was lucky enough to have parents who supported her love of mathematics, at a time when an interest in this subject was considered improper for a woman. In 1934, she became the first woman to be awarded a PhD in Mathematics at Yale University. In 1944, she was the third person to program one of the first electronic computers (the Mark I). She also virtually invented COBOL (the first computer language to use plain English). In fact, she was a trailblazer in the evolution of computers and their software, due to her pioneering work and leadership. She received the first computer science 'Man of the Year' Award from the Data Processing Management Association in 1969, was awarded some 47 honorary degrees, and received honours from two American Presidents before her death in 1992.

In 1945, she was working at Harvard on the Mark II Aiken Relay Computer. At 3.25 pm on 9 September, the operators started an adding test — but this primitive computer kept adding wrongly. So the operators went looking, and found a moth trapped between the metal contact points of relay 70, in panel F. At 3.45 pm, they had removed it from relay 70, panel F and sticky-taped it into the logbook, with the comment: '*First actual case of bug being found*.' Soon the story spread that the operators had 'debugged' the computer. Grace Hopper was not the person who actually found the moth, but, because of her senior position with the Mark II computer, and because of the respect that everybody had for her, she became associated with the term 'computer bug'.

Even with error-checking, there remain anywhere from 0.5–5 errors per 1000 lines of code. Think about that next time you fork over 'big bucks' for software.

All in a day's work

blame the programmer

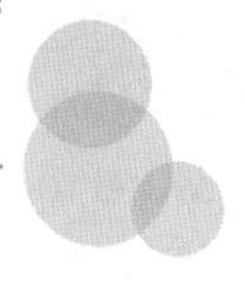

Today, we have the situation where software is released with between 0.5 and 5 bugs per 1,000 lines of code. Most of these bugs will never be a problem to most users. On 2 October 2002, Steve Ballmer, the Chief Executive Officer of Microsoft said that *'1% of the bugs in Microsoft Corporation's software cause half of all reported errors, and ... 20% of (the) bugs are responsible for 80% of the mistakes.'* Now this is one area where software engineering is different from other types of engineering. A structural engineer, who designs a bridge that will safely carry 5,000 tonnes of load, would be completely confident that the bridge could safely carry a smaller load of 500 tonnes. But you can't make that assumption in software engineering. A payroll application might handle salaries of up to $300,000 per year, but might crash if asked to process the specific salary of $33,333.33 per year.

Here's a recent example of a computer bug. I will mention Microsoft simply because they are the biggest software company in the world. On 25 October 2001, Microsoft released Windows XP operating system. That same day, they also released 18 megabytes of repair patches on their home page. Two of the many repair patches were supposed to fix some rather important security holes — and one of the patches worked, but the other one didn't!

To their credit, Microsoft has started thinking about how to solve the huge problem of computer bugs. In January 2002, Bill Gates proclaimed that all Microsoft employees should make '*reliable and secure computing their highest priority*', in an effort to dramatically reduce the number of bugs. In February, Gates stopped all new writing of software for two months. Microsoft programmers were brought, in groups of a thousand at a time, to a giant auditorium where *'embarrassing snippets of their flawed software'* were shown to their fellow programmers on huge screens. In May 2002, Microsoft helped found the US$30 million

Sustainable Computing Consortium at Carnegie Mellon University with NASA and 16 other corporations — with the aim of making software more reliable and less bug-prone.

Over the past few years, there have been massive internet problems with computer viruses and worms, like *Klez, Bugbear* and *I Love You*, to mention just a few. Will making software less bug-prone make it more resistant to viruses and worms?

Lovely Linux

Linux is a version of Unix, that ancient (but very reliable) operating system. Some of the young (and penniless) students in our School of Physics at the University of Sydney have picked up some ancient and slow laptops that were thrown out because they were too gutless to run any of the modern (and huge) operating systems. The students installed Linux onto these laptops, which then lets them do email, web-browsing, and word- and picture-processing, and anything else they want. They love Linux. Some of them have used their laptops for over a year without a single crash. Reliability is a good thing.

A version of Linux, Version 4.4, is at the core of the new Macintosh Operating System X (OS X). People tell me that one advantage is that when an application crashes, it does not affect any other application that happens to be running, and it does not crash the OS X. All they have to do is restart the single application that crashed. An unexpected benefit is that their productivity is now improved, because they are not wasting time waiting for the entire computer (operating system and all the other applications) to start up again.

so many bugs!

Computers are solid lumpy objects that we can touch, but the software that makes computers run is invisible. Software applications are just various arrangements of 1s and 0s, zipping around inside the dark and noiseless electronic corridors of the computer. The esteemed computer scientist, the late Edgser W. Dijkstra, was rather scathing about the quality of today's software. He said that the average user of computers *'has been served so poorly that he expects his system to crash all the time'*. This has given us *'a massive worldwide distribution of bug-ridden software for which'*, Dijkstra rightly said, *'we should be deeply ashamed'.*

There are a few reasons why software is so full of bugs. One is that software doesn't evolve by continual refinement, as most engineering does. In fact, engineers have a saying that *'form follows failure'*. Therefore, machines improve as the little failures get fixed. For example, the first model of any car usually has a few problems, but these get ironed out as the model evolves. And if you consider engineering over a longer time, today's cars are much more reliable, better finished, and have fewer nasties like noise and vibration, than the cars of the 1980s. But the ongoing reliability of software engineering is distinctly different from regular engineering. As each new version of software is nearly always very different from the previous version, there is no fine-tuning. Instead, each new version adds new problems that didn't exist before.

Customers are also to blame for the buggy software — they always want more features. This creates pressure on the programmers to make the software package do all things for all customers. So word-processing applications have exploded from one-quarter of a megabyte, to hundreds of megabytes — and yet, the average user wouldn't use more than a few percent of the program's potential.

Another reason is poor design, or even lack of design. For example, in some applications, simply placing the mouse pointer on the active window throws tens of thousands of unnecessary instructions at the computer's central processing unit, even though the program is doing nothing.

piling more into compilers

The philosophy behind writing software has also changed. Computers read only 1s and 0s, but software is written in plain language, or something like it (certainly not 1s and 0s). So special programs called 'compilers' turn the programmer's written instructions into 1s and 0s. In the 1960s, I would write my program, and then wait for a few days in the queue until my program ran on the big mainframe computer. If I made any mistakes, the compiler would refuse to run my program. If I wanted my program to work, I would have to fix the mistakes — or waste lots of time in the process. So in those days, the programmers would check their programs very carefully and try to remove all the mistakes before submitting the program to the compiler.

But the thinking is very different today. Because computers are everywhere, and relatively cheap, today's programmers use the compiler as the final checker. They write something, feed it to the compiler, fix the errors, feed it to the compiler again and fix the next batch of errors. They keep doing this until the program gets past the compiler. Of course, there are many errors that a compiler won't fix, such as errors in logic — and that's why, when you buy software today, there are about 0.5–5 errors per 1,000 lines of code.

Finally, there is the basic problem that the initial design for some large software projects is nothing more than some words and some bubbles, joined together by some straight and squiggly lines, scribbled on the back of an envelope.

zapping the bugs

There are many things that can be done to make software less buggy. First, software design meetings should involve all interested parties — from final customers to programmers, and administrators to business managers. Second, they should all have the same shared vision — but all too often, this changes midstream, for example, when a marketer tells the programmers that the customers really want some specific feature. Third, some programming languages, such as the military language called ADA, were designed so that it is impossible for the programmer to make certain mistakes. Perhaps new programs could be written using this philosophy.

But let's not be too negative about buggy software, because after all, it still works. In 1965, very little commercial software was available. By 1990, this market had grown to US$350 billion — and while software is a long way from perfect, it usually gets the job done. And that's kind of reassuring — maybe computers are like people, after all.

References

Buchholz, Werner, 'Comments, queries, and debate — origin of the term software: evidence from the JSTOR Electronic Journal Archive', *IEEE Annals of the History of Computing*, April/June 2000, pp. 69–71.

Mann, Charles C., 'Why software is so BAD', *Technology Review*, July/August 2002, pp. 33–38.

Roush, Wade, 'Writing Software Right', *Technology Review*, April 2003, pp. 26–27.

'Software bugs cost big bucks', *Science News*, 20 July 2002, p. 45.

Schwarz, John, 'Worm hits Microsoft, which ignored own advice', *The New York Times*, 28 January 2003.

You would think in the 21st century that the military with their substantial budgets would have only the latest and the fanciest technology flying in the sky. They do. Yet surprisingly, some of the machines run on propellers — made of wood!

wooden propeller

To an aeronautical engineer, a propeller is just a rotating wing. You can feel what air does to a wing if you recklessly put your hand out of the car window, while travelling at speed. If the tips of your fingers are angled upward, the air pushes your hand upward (with the opposite occurring, if your fingers are angled downward). The blades of a propeller cut through air or water. Because the blades push the air or water backward, the propeller moves forward. (By the way, in the United Kingdom, they call front-mounted propellers 'airscrews', because they 'screw' through the air, pulling the craft behind them.)

The amount of push, pull or thrust that a propeller generates depends on two factors — first, the mass of air or water that it's moving, and second, how quickly it accelerates through this mass. In general, really efficient propellers shift a lot of mass, but with relatively little acceleration.

The blades can be set to cut through the air at either a shallow or a steep angle. In theory, a 41-pitch propeller would move 41 inches (104 cm) forward through the air when the propeller blade had made one complete revolution. In the same way, a 50-pitch propeller would move forward 50 inches (127 cm) per revolution.

Fixed-Pitch Propeller

A fixed-pitch propeller is always a compromise.

For example, Formula 1 propellers are about 90% efficient at turning energy into forward motion — so these light and powerful air racers can zip along at up to 500 kph at 7,000 feet. The propeller makers need to know the engine power, the propeller diameter and the engine revolutions. But these props are very inefficient at low speeds — it's like having a car with only one gear.

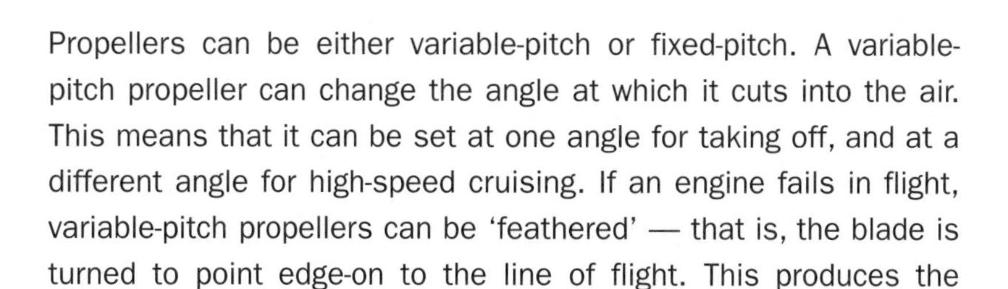

variable and fixed propellers

Propellers can be either variable-pitch or fixed-pitch. A variable-pitch propeller can change the angle at which it cuts into the air. This means that it can be set at one angle for taking off, and at a different angle for high-speed cruising. If an engine fails in flight, variable-pitch propellers can be 'feathered' — that is, the blade is turned to point edge-on to the line of flight. This produces the least drag, and the crippled plane can glide a long way. But

Come fly with me...

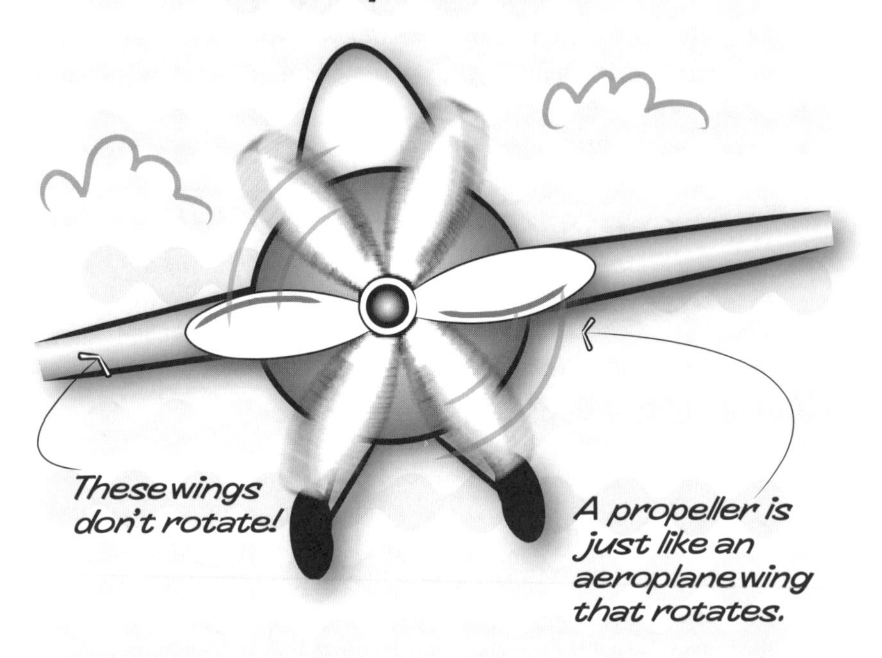

The blades of a propeller cut though air or water

Catching the cool breeze

variable-pitch propellers have a lot of complicated machinery in the hub, which makes them more expensive.

Fixed-pitch propellers may be cheaper, but they are always a compromise. A target-drone, designed to be shot at for training practice, usually comes with a fixed-pitch propeller. The fixed-pitch propeller of a target-drone would typically be rated at 90% efficiency when cruising at 480 kph. But the same propeller is so inefficient at low speed, that it can't get the target-drone off the ground. To take off, it actually has to be catapulted into the air.

sensenich propellers

Today, there are metal propellers and wooden propellers — in fact 10% of the total aviation market runs on wooden propellers, mostly Sensenich propellers.

The Sensenich propeller got its start on a summer day in 1928, when Martin and Harry Sensenich bolted an engine and a propeller to a farm wagon, and took a crazy high-speed ground run on the dirt roads of Lancaster County, in Pennsylvania. They created such a furore, including the stampeding of several herds of cattle, that they were banned from running their propeller-powered farm wagon on the roads.

But the boys weren't going to be stopped, and by winter they had harnessed their engine and propeller to an ice sled. They staked down a strong piece of wood in a nearby frozen river, and tied their propeller-driven ice sled to the wood with a 30-metre length of rope. They had lots of fun doing 60-metre circles until the rope broke, and they were thrown into the bushes on the frozen riverbank. They were able to pick themselves up, but their propeller had shattered into a thousand pieces. So they decided to make propellers, using tools that normally make wagon wheels — and three-quarters of a century later their company is still making propellers.

From those early beginnings, Sensenich has made around 450,000 wooden propellers. In the USA, Sensenich supplies almost all of the fixed-pitch wooden propellers that pass certification by the Federal Aviation Administration. Their factory, which employs 20 people, turns out about 4,000 fixed-pitch wooden propellers each year.

The Sensenich Company will also check, repair and rebuild wooden propellers. Some propellers are sent in for repair after two years, but because of bad maintenance in the field, they're thoroughly un-airworthy. In contrast, some 30-year-old propellers only need a new coat of lacquer.

from plank to propeller

Sensenich's propellers are made from rough-cut yellow birch planks from New England tree farms. All the planks are one inch (about 2.5 cm) thick and come in random widths and lengths. They are checked for any weakness, and then glued together into a stack with as many as 16 layers. Believe it or not, a stack of thin strips of wood glued together to make a 10-cm-thick lump of laminated wood is stronger than a single 10-cm-thick piece of wood. The planks are stuck together, using an ancient glue called 'Resorcinol' — it's the only glue that Sensenich has used for the past 50 years.

Immediately after gluing, the individual strips of timber are clamped for 24 hours to laminate them into a single slab. This slab is then rough-shaped on carving machines. Skilled workers do the final adjustments by hand, using a combination of power and hand tools. Stainless steel and/or brass is then applied to the twisting curve of the leading edge of each blade of the propeller — to make it more resistant to bird strikes. At every individual stage of production, the propeller blade is balanced again. In its finished form, a 3-metre-long propeller blade is so finely balanced that one paper clip placed on one of the tips will slowly rotate it out of balance. At full speed, the very tips of this propeller can bite into the air at around 980 kph — and thanks to the exquisite manufacturing process, can do so for thousands of hours without any vibrations.

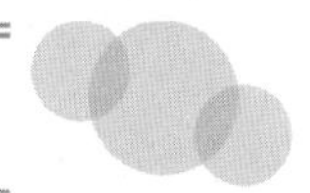

sensenich propeller users

Today, about 10% of Sensenich's propeller blades go to the display trade, where they are used for decoration only (e.g. on the wall of a private home or a fancy sports bar). Airboats use around 20% of their production. A typical airboat or swamp-boat can reach 110 kph in shallow water, because it's pushed by a four-blade, two-metre propeller mounted on a 400-kilowatt engine. About 40% of Sensenich's output goes into general light planes — often disparagingly called 'puddle jumpers'.

But the big surprise is that 30% of their propellers go into the latest Unmanned Aerial Vehicles (UAVs) — small reconnaissance and (recently) attack planes. These UAVs can cost US$1 million each. They can be launched from a truck or a ship, fly around for hours or days, and finally return to crash into a $20,000 Kevlar net. All that breaks is the $300 propeller.

Back in 1955, the US Air Force made the loudest aeroplane ever built.

Back then, a jet-fighter had a great top speed, but lousy acceleration. On the other hand, piston-propeller fighters had great acceleration, but lousy top speed.

Wood versus Metal

Maybe, they thought, if you put a propeller on the front of a jet-fighter, you could get both acceleration and top speed. So the engineers modified a Republic F-84 Thunderjet by coupling its huge jet engines via two driveshafts to a single gearbox in the nose, which then spun a single propeller. It was now labelled the XF-84H.

The outer tips of the propeller were travelling faster than the speed of sound, so they made a horrendous racket. The Thunderjet was so loud that it could be heard 22 miles away (that's about 35 km!). Maybe that's why they called the Thunderjet the Thunderscreech.

wood is stronger than metal

Wooden propellers can only have a fixed pitch, because they are fashioned as a one-piece entity along with the hub. Metal propellers can have variable pitch, but believe it or not, wood still has advantages over metal. Why?

For one thing, wooden propellers deal with the many vibrations of the engine and the aeroplane body thousands of time better than metal propellers. Being bent backwards and forwards millions of times makes a metal propeller build up invisible internal flaws — but wood is not affected by this vibration cycle. With a metal propeller, you can't see the damage because it's hidden. But a wooden propeller is a good case of *'what you see is what you get'*.

Because metal is stronger than wood, you would think that a metal propeller could be made thinner and more efficient. However, in the 1950s, aeronautical engineers came up with a new design for wooden propellers that brought their performance to within a few percentage points of the efficiency of metal propellers.

Wooden propellers have a further advantage. They can be bolted onto engines that run from 2,700–6,000 rpm — which can't be done safely with a metal propeller. And of course, you can easily carve a new wooden propeller for a few hundred dollars but it costs thousands to cast, grind and then polish a metal propeller.

So even in the 21st century we have a new spin on propellers, because wood still works.

References

Encyclopaedia Britannica, DVD, Encyclopaedia Britannica, 2003.

Harpole, Tom, 'Good Wood', *Air & Space*, June/July 2003, pp. 64– 1.

'ZZWRRWWWBRZRR — How the XF-84H Thunderscreech broke the noise barrier', *Air & Space*, June/July 2003, pp. 56–61.

The 'Mexican Wave' first became famous during the 1986 soccer World Cup in Mexico. In fact, that's how the Mexican Wave got its name, because it got its first worldwide exposure at this event — and soon enough, there was a swell of interest in far-away Europe.

mexican wave

If you have ever been to, or watched a major sporting event, you'll probably have seen the famous Mexican Wave. This wave, which sweeps around the audience in a stadium or sports arena, begins when one group of people leap to their feet with their arms up and then sit down. The group of people next to them then do the same thing, and so on. If you're on the other side of the stadium looking across, you can see this beautifully rhythmic and synchronised movement rolling through the audience. On a good night, you can see multiple waves winding their way around the terraces. Some scientists have studied this strange phenomenon, and not only do they now understand it, they can probably do something useful with this knowledge.

The Wave

In North America, the Mexican Wave is just called 'The Wave'. There are two North American claims to having invented 'The Wave', but they refer to events years later, in October 1981. One claim is that 'Krazy' George Henderson invented it at a baseball match — Oakland *vs* New York — on 15 October 1981.

The other claim is that the Washington Huskies, a college football team, first performed it on 31 October 1981. That particular primordial wave was led by the University of Washington Husky 'yell-leader' Rob Weller, and another 'yell-leader' from the University of Washington, the marching band director, Bill Bissell. Of course, other people claim that the Mexican Wave started when the very first Mexican left home.

so how does a wave start?

Scientific work was done by Támas Vicsek and his colleagues from the Eötvös University in Budapest in Hungary and the University of Technology in Dresden in Germany. They videotaped, and then analysed, 14 separate Mexican Waves in football stadia each holding more than 50,000 people. They noticed that the wave generally went in a clockwise direction — that it spread from one person to the next person on their left. The Mexican

Wave usually moved at around 12 metres (or 20 seats) per second, and was about 9 metres (about 15 seats) wide.

excitable media

Mathematicians have made mathematical models of much of the world around us, with some of them creating mathematical models of what are called ‘excitable media’. One example of ‘excitable media’ is the dry trees and dry litter in a forest. This particular theory of ‘excitable media’ describes how a forest fire starts, and then spreads.

Another example of ‘excitable media’ is the set of muscles that make up your heart tissue. When they are given a little jolt of electricity, the muscles in your heart will contract. It turns out that there are quite a few electrical abnormalities of the heart that involve problems with the excitability of the heart tissue — either it's too excitable, or it's not excitable enough. In fact, the theory of ‘excitable media’ can also predict that some people will have extra heartbeats.

But it was a bit of a surprise to find that this theory of ‘excitable media’ could deal with something as complicated as human behaviour — but then again, we are talking about a sports event.

You can think of a person as being an excitable unit — and that's not so unusual. This simply means that a person can be prompted into action by some sort of external stimulus. In general, the closer and more powerful the stimulus is, the more the human being, as an excitable unit, will respond. So to explain how a Mexican Wave can get started and keep rolling, each person needs only three internal rules that they obey, one after the other. First, they wait in their resting state, ready to be excited by the right stimulus. Second, when stimulated, they go through the active phase where they stand up and wave their arms. Third, they progress to the refractory phase, which is what you and I would call ‘sitting down again’.

Eager 'Mexican Wavers' in an excitable unit

You now might say that this research is incredibly useless. But it is important with regard to crowd control. When you have 100,000 people at a sports event, simply moving them in and out of the stadium can be dangerous if it's not done wisely. Even something as simple as putting a handrail along the centre of a corridor can speed up the movement of people in or out of a stadium. So knowing this theory of excitable media could help if a small group of agitators tried to get a large crowd over-excited.

a wave to pass the time

The scientists did find a good relationship between their theory, and what they actually saw in real life. Basically, you need a minimum critical mass of two or three dozen people to get the wave going. And even that's not enough — there needs to be a lull in the sports event. After all, if the football game or athletics event is incredibly riveting, the audience is not going to pay attention when the person next to them suddenly jumps up.

So if you do see, or are involved in, a Mexican Wave, this research also tells you that you're not getting your money's worth — because the game is boring, and the crowd are entertaining themselves.

References

Fountain, Henry, 'The physics of the wave, stadiums, not ocean', *The New York Times*, 17 September 2002.

Vicsek, T., Farkas, I. & Helbing, D., 'Mexican waves in an excitable medium', *Nature*, vol. 419, 12 September 2002, pp. 131–132.

Champagne has been a celebratory drink for centuries. But what makes the bubbles much more special in champagne than in other fizzy drinks? And do the bubbles really make you more tipsy than when drinking still wine?

beautiful bubbles

Humans have grown to love a fungus called *Saccharomyces cerevisiae* — brewer's yeast. It has given us beer for thousands of years, and champagne for centuries.

But it's only been in the 21st century that we have begun to understand how bubbles form in champagne.

bubbly makes you tipsy

People say that champagne *'goes to straight to my head'* — and science has proved that they're right.

Fran Ridout and her team from the Human Psychopharmacology Unit at the University of Surrey studied 12 volunteers to see how champagne 'bubbles' affected their blood alcohol level. In the study, the volunteers drank two versions of champagne — freshly opened fizzy champagne bursting with bubbles, and flat champagne that had been criminally de-bubbled with a whisk.

In each session, the volunteer drank only two glasses of either bubbly or non-bubbly champagne. The amount of champagne given to each person was adjusted according to their weight, so that they drank the same amount of alcohol per kilogram of body mass.

After five minutes, the fizzy champagne had pushed the blood alcohol level to 0.54 mg of alcohol per ml of blood (just over the legal limit for driving in Australia), while the flat-champagne drinkers averaged only 0.39 mg/ml. After a further 35 minutes, fizzy drinkers were up to 0.70 mg/ml, while flat-champagne drinkers still trailed behind at 0.58 mg/ml. So there's something about the bubbles in champagne that makes you more intoxicated.

The higher blood alcohol levels affected the bubble drinkers in two ways.

First, the bubble drinkers took an extra 200 milliseconds to notice an object, as compared to when they were sober. The flat-champagne drinkers took only an extra 50 milliseconds. Second, the bubbly champagne drinkers had more trouble in spotting which numbers in a long list were odd or even.

why bubbles make you tipsy

So what's going on? Well, we do know some of the physiology of how molecules of alcohol get into your bloodstream by leaving your intestines, travelling through the gut wall and into a blood vessel.

It all begins with the alcoholic drink gurgling down from your mouth to your stomach. On average, about 20% of the alcohol gets absorbed in the stomach. The remaining 80% of the alcohol you drink gets absorbed further along in your intestines. But this depends on what food is already present in your stomach.

Sugars and fatty foods tend to close the outflow valve of the stomach. Alcohol passes into the bloodstream slowly from the stomach. If the alcohol remains in the stomach longer, this causes a slow increase in blood alcohol levels. However, if the stomach is empty, the alcohol quickly moves on into the small intestine. The small intestine has a thinner wall structure, so the alcohol passes through the wall rapidly, which increases blood alcohol levels faster.

Even today, we still don't know the exact pathway of where (and how) the carbon dioxide in champagne bubbles speeds up the absorption of alcohol.

Perhaps someone could do a study with beer (which also has bubbles of carbon dioxide) or Guinness, which has bubbles of nitrogen.

Champagne Burn

The 'champagne burn' is one of the delights of drinking champagne. One cause of the 'burn' is the splatter of droplets of champagne landing on pain and touch receptors in your nose and tongue.

But Earl Carstens, from the University of California at Davis has another theory about the 'champagne burn'. He reckons that carbonic acid on your tongue also causes some of it.

Carstens used a drug called Acetazolamide. Acetazolamide stops carbon dioxide being turned into carbonic acid (it's often used to combat altitude sickness). People have long noted that one of the side effects of this drug is that it lessens the tingle that you normally get from fizzy carbonated drinks.

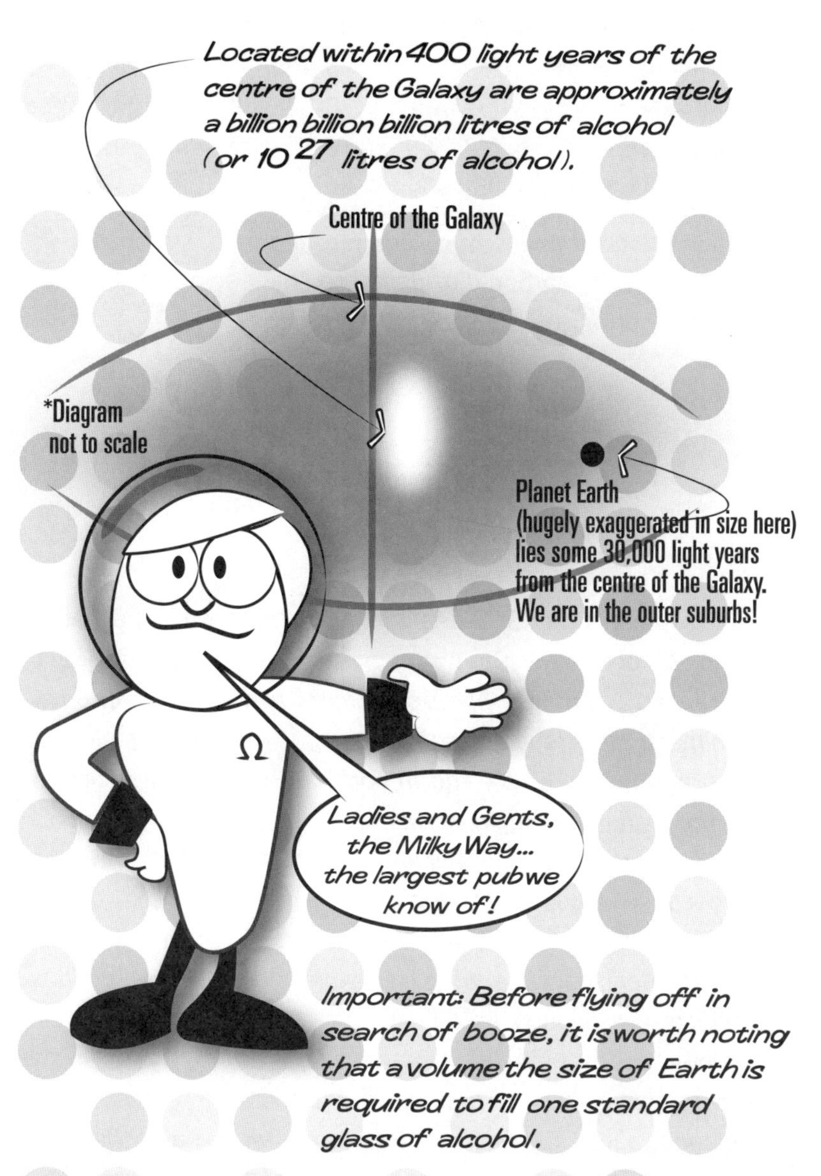

Space, the ultimate pub crawl

Carstens painted half of the tongue of each of his volunteers with Acetazolamide. He then asked them to dip their tongues into various test drinks, and report on a scale from 0–10 how irritating or burning the sensation was. Zero was no sensation, whilst 10 was the rating given to hot peppers. The scores were lower on the side of the tongue that had been coated with Acetazolamide. If carbonic acid was not formed, there was less burning sensation.

So in the champagne burn, some of the burn is probably caused by the carbon dioxide in the liquid being turned into slightly corrosive carbonic acid.

alcohol — good and bad

Alcohol is tricky stuff. Shakespeare got it right when he had the porter in *Macbeth* say: *'it provokes the desire, but it takes away the performance ... it makes him and it mars him; it sets him on and takes him off; it persuades him and disheartens him; makes him stand to and not stand to; and in conclusion equivocates him in a sleep and, giving him the lie, leaves him.'*

Shakespeare was scientifically correct in talking only about men, because alcohol has different effects on men than on women. Women can't metabolise (or break down) alcohol as quickly as men, because a key enzyme, alcohol dehydrogenase, doesn't work as well in women.

Alcohol causes temporary impotence (Brewer's Droop) in men for two reasons. First, it reduces the amount of testosterone in the bloodstream. Second, it interferes with nerve impulses controlling the muscles in the penis that allow an erection to occur. But in women, low doses of alcohol will temporarily increase the blood testosterone level — and this increases sexual desire and libido.

On the other hand, long-term moderate alcohol consumption does interfere with women's fertility — as Mary Anne Emanuele from Loyola University's Stritch School of Medicine discovered in her study. She found that 50% of women who had three drinks a day either ovulated late, or not at all. The same occurred in 60% of women having eight drinks a day.

In small doses, though, alcohol can be good for you.

god loves alcohol

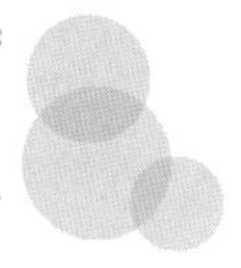

Alcohol has been part of our culture for thousands of years. It's a wonderful social lubricant.

Ignore the cold heartless words of the World Bank, which defines ethyl alcohol as *'a colourless, flammable liquid used to preserve fish'*. Benjamin Franklin came much closer to the truth when he said: *'Beer is proof that God loves us, and wants us to be happy.'*

After all, God (who is more important than the World Bank) must love alcohol also, because He gave us a huge cloud of alcohol near the centre of our galaxy, the Milky Way.

brewery at heart of galaxy

The Milky Way is about 100,000 light years across. Our solar system is out in the suburbs — about 30,000 years from the centre. But near the centre, about 400 light years from the Black Hole at the heart of our galaxy, is a giant 'molecular cloud' called Sagittarius B2 — and it's full of alcohol.

Stars do nuclear burning of hydrogen to make heavier elements such as helium, carbon and oxygen. This nuclear burning releases huge amounts of energy. As the star evolves, it squirts out these elements into space. These atoms gather together under the force of gravity, and form 'clouds' such as Sagittarius B2.

Sagittarius B2 is huge — about 150 light years across. It makes up about 10% of the visible mass of our galaxy. It has regions called 'hot molecular cores'. In these regions, new stars are born — and various molecules, including alcohol, are made from atoms.

Sagittarius B2 holds about a billion billion billion litres of alcohol, but B2 does not rain alcohol in big drops. Instead the alcohol is spread so thin that if you wanted to collect a glass of it, you would have to trawl from a volume as large as our planet. (But we Australians have been known to go to great lengths to get a drink.)

Champagne Fog

When you open some chilled champagne (or even a lesser fizzy drink), you'll sometimes notice a thin fog at the mouth of the bottle.

When you open the bottle, the gas expands rapidly. It has to work as it pushes against the atmosphere outside. That work needs energy, which comes from the internal energy of the expanding gas. If you take energy out of anything, you reduce its temperature.

But what made the cold expanding gas turn into a fog?

To understand this, you need to know that water molecules (H_2O) are in the shape of a little boomerang. The single oxygen atom is in the middle of the boomerang, while the two hydrogen atoms on the tips. These boomerang-shaped molecules are always running into each other.

If the temperature is high enough, they have so much energy that as soon as they hit each other they immediately bounce off again. But when the temperature is lower, they have less energy, and so they can stick to each other. When enough water molecules stick together, they make up tiny droplets of water that 'float' in the air on the wind currents. If there are enough of these droplets to block your vision, you now have a fog.

You don't need a lot of water vapour to make up a thick fog. A fog can be so thick that you can't see a car 10 metres in front of you — and yet you'll need to use the windscreen wipers as little as once every kilometre. This shows that you're not picking up a lot of water on the windscreen. But even this tiny amount of water, if it is divided finely enough, can reduce your vision to only 10 metres.

yeast makes alcohol

Brewer's yeast also makes alcohol. This yeast is a bit like us. It loves to eat sugars, and it gives off carbon dioxide as a waste product. But unlike us, brewer's yeast makes another waste product — alcohol.

Humans tamed brewer's yeast a long time ago. When archaeologists studied ancient pottery from a small Neolithic village called Hajji Firuz Tepe, in the Zagros Mountains of Iran, they found the chemical signatures of various alcoholic drinks. So 7,000 years ago, in Iran, our ancestors were fermenting grapes.

The Ancient Egyptians have left us 4,500-year-old written records in which they describe how grapes are used to make wine. In the Old Testament many references are made to wine. Over 2,000 years ago, both the Greeks and the Romans planted grapes in any of their colonies that were suitable. The Catholic Church incorporated wine into its mass, and in many of the religious orders monks grew and processed grapes as part of their daily routine.

Much more recently, in 1698, a 60-year-old monk, Dom Pierre Perignon, invented sparkling champagne. You make wine by adding yeast to wine. His brilliant concept was to throw in another step. He added extra yeast and sugar to wine, then left it to ferment in the bottle to create more carbon dioxide. He reportedly said to his fellow monks after tasting the sparkling wine for the first time, *'Come quickly, I'm drinking stars!'*

Yeast/Fungus and Alcohol — The Beginning

The fungus we love so much is called brewer's yeast. Its fancy name is *Saccharomyces cerevisiae* — which literally means 'sugar mould of beer'.

S. cerevisiae is all over our planet — floating in the air and sitting on surfaces. The first lucky accident probably happened about 10,000 years ago when some humans left out a bowl of a sweet food (dates or honey) mixed with water. Some brewer's yeast floated in and turned the sugars into carbon dioxide, ethanol — and no other nasty by-products.

This friendly fungus had a sweet tooth, and we probably quickly learnt how to use it.

Sweeter grapes with their higher sugar content were a better food supply than ordinary grapes.

Most grains have their sugars locked up in starch. But grains that have sprouted have turned the starch into sugar, making them a better food supply for the yeast.

Rice starch is fairly hard to turn into sugars. So the Japanese cultivated a different yeast, the Koji Mould called *Aspergillus Aoryzae*, that could do this conversion. Of course, once the brewers had the rice sugars, our friend *S. cerevisiae* turned the sugars into sake.

essentials of champagne

There are only three grape varieties allowed in the production of champagne — Chardonnay (a white grape), Pinot Noir and Pinot Meunier (both dark grapes). These grapes love the poor chalky soils of the Champagne region in France — north-east of Paris.

Another essential ingredient of champagne is its toasty flavour. This flavour comes from the chemical breakdown products of dead yeast cells.

And of course, there are the bubbles.

harvesting in champagne

Harvesting grapes in the Champagne region is the cultural climax of the year. Each village spends about 10 days harvesting. It's done entirely by hand, and during this short period, the Champagne region enlists the use of an extra 60,000 pairs of arms. The picked grapes are then carefully selected, and delivered to some 2,000 pressing houses scattered across the region.

The grapes are pressed to make a juice. Depending on the quality of the grapes in that year, they may be blended with grapes from previous years. To make white wine, the wine makers put the grape juice into an open vat and add yeast. (To make red wine, they add the skins of the grapes.) The yeast turns the sugars into carbon dioxide (which escapes into the atmosphere) and alcohol.

The juice ferments and bubbles for two to three weeks, giving a clear wine with no bubbles. The wines are then bottled and laid down in the cellars, that the locals call 'caves'. For instance, Möet & Chandon has over 28 km of caves, where the temperature is always between 10°C and 12°C and the humidity is high.

Foamy Champagne

When you pour champagne, you get a bubbly mousse of foam. This is very important to wine tasters, because the first contact between them and the wine is sight. This bubbly frothy foam then disappears quickly, to be replaced by a collar of rising bubbles. This collar survives for quite a while because of certain chemicals, which were discovered by Michel Valade and his team at the Comit Interprofessionnel du Vin de Champagne in Epernay in France. They repeatedly filtered the champagne, to remove successively smaller molecules. They reported that the molecules that keep the bubbles surviving as long as they do are proteins and polysaccharides, with a molecular weight between 10,000 and 100,000.

In general there are two types of foam — 'wet foam' and 'dry foam'.

Champagne has 'wet foam', where the bubbles are little spheres that are touching each other. Beer has 'dry foam', where any liquid drains back quickly into the beer leaving behind bubbles with many sides — polyhedrons.

'Foamability' refers to how much foam is created when champagne is first poured into a glass.

'Foam performance' tells you how well the champagne creates a ring or collar inside the glass.

'Foam stability' describes how long that ring lasts for.

méthode champenoise — 1

The essential part of the *Méthode Champenoise* is that you add yeast and sugar to the wine — and then age the wine for several years with the yeast inside the capped bottle. The yeast cell multiply so rapidly that after just one week, there are about 10 million yeast cells per millilitre.

But after five to eight weeks, the fermentation (and the manufacture of alcohol and carbon dioxide) stops, and the yeast begins to sink to the bottom of the bottle. The carbon dioxide is trapped in the bottle, and so it is absorbed into the wine.

The bottles are then laid down for a few more years on their sides — called 'ageing on lees'. During this stage the dying yeast cells release enzymes, which will destroy the yeast cells (nature tidies up after itself — isn't that sweet?). The disintegrating cells then release a veritable galaxy of different chemicals (polypeptides, nucleic acids, polysaccharides, etc.) into the wine. These chemicals influence the final aroma, taste and bubbles of the champagne.

méthode champenoise — 2

The process that removes the yeast sediment is called 'riddling'.

The bottles are aged with the mouth of the bottle lower than the base, so that any sediment collects on the inside of the cap. Over the next two months, each bottle is tilted a little more each day until it is completely upside down with the cap lowermost.

The bottle is also twisted a little each day, so that the tipping-and-twisting works the sediment down to the neck.

The sediment is then removed by a clever trick. Salty water at –25°C is placed around the neck, and freezes the champagne into a solid frozen band. The cap is removed, and the sediment is drained.

The next step is the 'dosage'. Sugar syrup dissolved in wine is added to the bottles. Depending on how much sugar is added, the final champagne is Brut (very dry), Demi-sec (quite sweet) or Sec (sweet and full).

Finally, a cork is inserted and the wine comes back to the caves for another rest.

vintage champagne

In most years, champagne is made by blending up to 40 different wines of various varieties and vintages.

But every now and then, a Vintage Wine is declared. All the still wines that are blended for that particular Vintage Year must come from grapes of that year's harvest. The wine makers also put aside at least 20% of that year's still wine, to be blended into the wines of future years.

Vintages are aged for four to eight years before they are sold. This lets the aromas develop complexity and depth. These aromas reflect the conditions of the grapes when they were harvested, such as their sugar content and acidity.

Every annual harvest is different, and always causes worry and concern.

For example, in 1993 the grapes got off to a good start with a fair winter and a mild spring, leading to some extremely promising grapes as harvest approached. Unfortunately, a very intense rainfall right at the beginning of the harvest slowed the full ripening of the grapes, so only some of the grapes were good enough to be picked.

The bubbles have gone straight to my head...

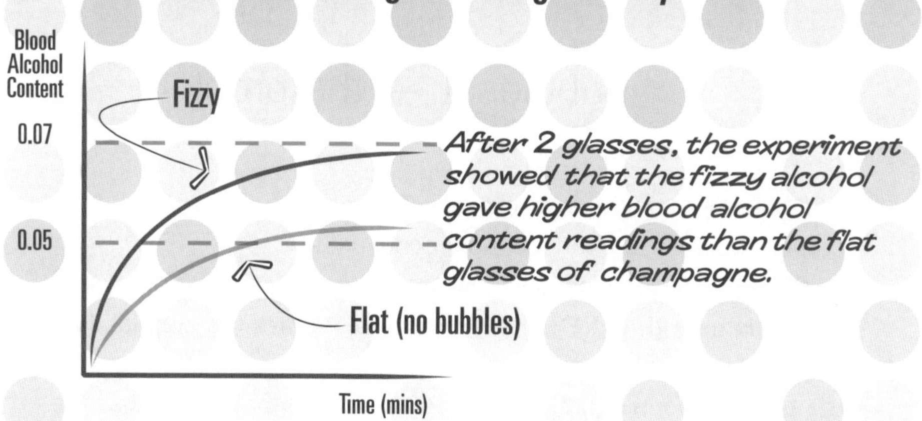

On the other hand, 1982 was a dream year for champagne. A cool winter (without frost) was followed by a long, slow, warm summer, and topped off with ideal weather during harvest.

Champagne Storage

Champagne does not improve with age significantly, and is ready to drink when you buy it. But you can cellar it for another four or five years (8–10 years for vintage champagnes). But you must store it horizontally, to stop the cork from drying out.

Champagne best releases its aromas at temperatures around 8–10°C. The experts say that you should never put champagne in the freezer to give it a quick chill, because in a few minutes, you can destroy years of work in trying to make the magnificent bubbles.

bubbles 1 — the problem

A bubble in champagne has three stages in its short life — it is born, it floats to the surface, and it collapses.

Champagne bubbles have always bothered scientists. They had no problems with the Physics of Bubbles — after the bubbles had popped into existence. But scientists had lots of problems with the bubbles popping into existence. Physicists have said that the pressure inside a bubble just after it's formed is incredibly high.

How could the carbon dioxide dissolved in the liquid wine push its way into the baby bubble, against such a high head of pressure?

Gérard Liger-Belair, a Professor of Physical Chemistry, has studied this 'incredibly important problem' for seven years. As a consultant for Möet & Chandon, he is such a physical chemist expert that he calls champagne wines 'multicomponent hydroalcoholic systems'. In his scientific papers, he talks about 'standard commercial champagne wine'.

His theory is that the bubbles were helped into existence by tiny hollow cylinders of cellulose — probably cast-off fibres of paper or cotton. His theory solves the 'big problem' caused by Laplace's Law.

bubbles 2 — physics of the problem

The genius, Pierre-Simon marquis de Laplace (1749–1827), was a mathematician, physicist and astronomer. Besides doing some of the first science on black holes, and the stability of the solar system, he also studied bubbles.

He explained what happens when you try to blow up a balloon.

You've probably noticed that it's very hard work to blow up a balloon when it is very small. Sometimes, it's such hard work that it's actually impossible for you to get the balloon started — no matter how hard you blow. You may have to get another person who can generate more pressure in their mouth. But once you get over the hard bit and force the balloon open, the work gets easier as the balloon gets bigger.

This is Laplace's Law — the smaller the bubble, the bigger the pressure inside. (The opposite is also true — the bigger the bubble, the smaller the pressure.) So if the bubble is just a few atoms across, the pressure is incredibly high — thousands of atmospheres. For comparison, the pressure in car tyres is only a few atmospheres.

bubbles 3 — roll down the hill

In nature, stuff 'naturally' moves from one place to another, if it can move, roll, or fall down some kind of 'energy hill'.

A ball will naturally roll down (not up) a hill — that's a 'gravity hill'. Heat will naturally flow down a 'temperature hill' — from hot to cold. It never flows from cold to hot, unless you push it with energy (such as in a refrigerator). And gas naturally moves down a 'pressure hill' — from high pressure to low pressure. You can see this in the weather, when 'wind' moves air from regions of high pressure to regions of low pressure. You can also see this in a car tyre, when air moves from the higher pressure in the pump into the lower pressure inside the tyre.

Now think about a liquid with molecules of gas in two places — first, inside a bubble, and second, dissolved in the liquid. The molecules will move only from the high pressure zone to the low pressure zone.

If the gas dissolved in the liquid wants to get inside the bubble, it has to be at a higher pressure than the inside of the bubble. It's very hard to get a pressure (in the liquid) that's higher than 'thousands of atmospheres' (which is the pressure inside a very small bubble). That's why it's very hard to start a very small bubble.

This was the 'big problem' that scientists had with champagne bubbles. How does a bubble get started if it takes enormous pressure to force gas into it?

bubbles 4 — the forbidden zone

Let's start with the fact that carbon dioxide dissolved in champagne is at a pressure of about 6 atmospheres. If the bubble is zero size, the pressure inside is infinite. So the carbon dioxide cannot leave the liquid to get inside a really small bubble — because 6 atmospheres of pressure cannot overcome infinite pressure.

But let's look at a bigger bubble, say 0.2 microns in diameter. Laplace's Law tells us that the pressure inside a bubble 0.2 microns in diameter is about 12 atmospheres. Once again, the 6 atmospheres of pressure inside the champagne cannot overcome 12 atmospheres of pressure.

The break-even point is around 0.4 microns. This corresponds to a pressure of 6 atmospheres. The carbon dioxide still can't get out of the liquid because the pressure inside and outside the bubble is the same.

It all changes when the bubble is bigger than 0.4 microns. The pressure inside is now less than 6 atmospheres, and the pressure outside the bubble (6 atmospheres) is greater than the pressure inside the bubble. The gas dissolved in the champagne can force its way into the bubble.

So how do champagne bubbles form, if you can't have bubbles smaller than 0.4 microns? How do you cross the 'forbidden zone' between 0 and 0.4 microns?

bubbles 5 — mystery solved

The old theory claimed (without any real proof) that there were tiny rough spots on the surface of the glass. When you poured champagne into the glass, these rough spots would supposedly trap little bubbles of air. These rough spots (and the bubbles they formed as the champagne filled the glass) were assumed to be bigger than 0.4 microns. This meant that you never had a bubble in the 'forbidden zone' between 0 and 0.4 microns. There was one major problem with this theory — when scientists went looking, they found the rough spots were very much smaller than 0.4 of a micron.

But Professor Liger-Belair has almost certainly solved the problem of the 'forbidden zone'. He has studied the bubbles in all phases of their life with high-speed photography. He looked very carefully at 'bubble nurseries' — where bubbles pop into existence. He found that they are impurities stuck to the glass wall — typically, hollow, roughly cylindrical fibres of cellulose. These hollow fibres are bigger than 0.4 microns on the inside.

When champagne is poured into the glass, it can't fully wet the inside of a hollow fibre, so the air remains inside. A typical hollow fibre is filled with a long cylindrical bubble bigger than 0.4 microns across. The pressure inside is less than 6 atmospheres. The molecules of carbon dioxide dissolved in the champagne are at a pressure of 6 atmospheres. This is high enough for the carbon dioxide to force its way into these bubbles of air.

bubbles 6 — bubbles grow

The carbon dioxide then makes the bubbles grow bigger, and as they get bigger, the pressure gets less. This makes it easier for even more carbon dioxide in the liquid (at 6 atmospheres) to force its way in. Eventually the bubble grows so large, that its buoyancy makes it break loose and it rises.

But there's still a bubble left behind, trapped in the hollow fibre. Again more carbon dioxide enters from the liquid, generating another bubble. And so the process continues, with thousands of bubbles being generated inside the drinking glass.

You don't get many bubbles inside a freshly opened bottle of champagne, because the long storage of several years has thoroughly wetted nearly every available surface and hollow fibre inside the bottle.

Champagne has roughly three times as much carbon dioxide as beer. Champagne's bubble nurseries can give birth to about 30 bubbles per second, compared to only 10 bubbles per second for beer bubble nurseries. It turns out that bubble nurseries can squirt out bubbles at frequencies ranging from 1 per second, up to

30 per second. There is wide range of frequencies, because there's a huge range in the shapes, sizes and numbers of particles stuck to the glass.

Explosive Champagne

Champagne came into existence because of Dom Perignon's Flash of Genius to ferment the still wine again — and because he had access to the recently-invented thicker and stronger English glass.

You might have noticed when you last picked up a case of champagne that it's a lot heavier than a case of wine. It's heavier for two reasons.

First, there's extra carbon dioxide in each bottle, thanks to the extra fermentation step. The weight of this extra carbon dioxide is around 12 grams per litre. This means that a case of 12 standard bottles of champagne has about 108 grams of carbon dioxide, as compared to a case of wine which has. But that's not the main reason for the extra weight.

Second, the glass of a champagne bottle is much thicker than the glass of a wine bottle. The carbon dioxide in the champagne bottle at the end of fermentation has a pressure of around 6 atmospheres — this works out to 60 tonnes per square metre. So the glass has to be extra thick and strong (and heavier), to contain the 6 atmospheres of pressure.

This pressure of 6 atmospheres will occasionally catastrophically explode the enormous Balthazar (equal to 16 bottles) and Nebuchadnezzar bottles (equal to 20 bottles).

bubbles 7 — speed of rising bubbles

Champagne bubbles rise more rapidly than beer bubbles. The study of rising bubbles goes back to the early 16th century, when Leonardo da Vinci investigated how fast bubbles of different sizes rose in various liquids.

Once again, Professor Gérard Liger-Belair has given us the answer.

He also found out why bubbles rise more quickly in champagne than in beer — even though they both contain alcohol and carbon dioxide.

First, there is roughly three times more dissolved carbon dioxide in champagne than in beer. As a champagne or beer bubble rises, it gets bigger (because the pressure of the surrounding liquid gets less). As it gets bigger, its internal pressure drops (because of Laplace's Law). This means that it will absorb more gas from the liquid into the bubble. But there's three times more carbon dioxide available to enter the bubbles in champagne, as compared to beer. So champagne bubbles will expand more rapidly than beer bubbles, and have more buoyancy, and therefore will rise faster.

Second, beer contains about 30 times more proteins and glycoproteins than champagne — and these chemicals slow down the rising beer bubble. These chemicals are called surfactants (short for 'surface active agent'). A surfactant is a chemical with two ends — one end dissolves in water, while the other end does not. So the end that dissolves in water will dissolve in the wall of the bubble, while the other end sticks out into the surrounding liquid. As the bubble rises, the bits that stick out slow down the rise of the bubble. Beer has 30 times more surfactants, so it has lots more prongy bits sticking out of the bubble, which makes the bubbles rise more slowly. These surfactants also slow down the rising bubble for another reason — they make the bubble more rigid, which creates more viscous drag on the rising bubble.

bubbles 8 — the surface

The surface of the foaming champagne in the drinking glass is made of thousands of bubbles. They pop into and out of existence so rapidly that we can see them only with high-speed photography, such as that used by Professor Liger-Belair. As the bubble reaches the surface, it actually bulges slightly above the surrounding surface of the champagne.

Thanks to gravity, the champagne liquid then drains out of the walls of the bubble down to the level of the surrounding surface. Once the wall gets too thin, it ruptures, making a hole. This hole then spreads and makes every part of the bubble above the surface vanish very rapidly — within 10–100 microseconds. Suddenly, in the surface of the champagne there is an open cavity that is a few millimetres across — and it immediately collapses. As the in-rushing liquid meets in the middle of this previously-hollow space, it ejects a high-speed jet of liquid vertically above the surface.

This vertical jet breaks up into many small droplets, which are launched at speeds of several metres per second. But because the droplets are so small, they slow down rapidly, and travel only a few centimetres. But a few centimetres is far enough for them to land on the pain receptors in the nose, and the touch receptors in the mouth. The slightly acid content of the champagne gives a minor pleasant burning sensation — the famous 'champagne burn'. (By the way, if you wear glasses and put your face near freshly poured champagne, you can see the 'splatter' forming on your glasses.)

These bubbles bursting at the surface make up part of that unique and deliciously characteristic 'mouth feel' of champagne — which is an important part of drinking it. The bursting bubbles at the surface also spray out a fine aerosol or cloud of concentrated aromas and flavours which somehow contribute to the experience.

At any given moment the surface has dozens of short-lived, flower-shaped structures (collections of collapsing bubbles) winking in and out of existence. What a shame our vision is too slow to be able to see them with the naked eye.

Bubbles and Spoon

We all know what to do with a full bottle of champagne, but what do you do with a half-full bottle? The myth goes that you can keep the champagne bubbly by putting the handle of a spoon (or fork) in the mouth of the open bottle.

Various scientists, including the editors of *New Scientist* and contributors to my home page, have examined this myth. They have 'done the experiment'. They opened two bottles of champagne at the same time, drank some from each bottle, and then put them back in the fridge — one with a spoon in its mouth, and one without.

Then they have drunk the two champagnes, spread over a four-day period, and tested the amount of bubbles via the simple mouth-feel taste test. One ingenious person tested the amount of bubbles by tying a condom over the mouth of the two champagne bottles. He then shook them, and measured how much gas was generated by seeing how far the condoms expanded. The results were clear — putting a spoon in the mouth of an open bottle of champagne has absolutely no effect on the complex life and death of a champagne bubble.

How did this myth arise? Colas and lemonades have no alcohol, beer has about 5% alcohol and champagne has about

12% alcohol. The alcohol makes the liquid more viscous — a little like honey — and this naturally tends to trap the bubbles better than the low- or zero-alcohol drinks. So colas go flat quickly, while champagne keeps its bubbles best of all.

Possibly somebody, after a good night of drinking champagne accidentally left a spoon in the mouth of the bottle — and when they found that the champagne still had bubbles the next day, decided that the spoon had kept the champagne bubbly.

But the true answer comes only from experimentation, so I suggest that when you next have champagne, open two bottles — and have a spoon or fork handy.

special sound of champagne

A lot of energy is given off during the collapse of a bubble. Some of this energy goes into squirting a jet of liquid upwards, but some is stored in the thin liquid layer that makes up the surface of nearby bubbles. This extra stress makes the surrounding bubbles burst sooner than would be expected — in an 'avalanche' effect. This gives champagne a very unusual sound. That special fizzing noise of the bubbles bursting in the foamy head of champagne being freshly poured is not ordinary noise. According to Nicolas Vanderwalle from the University of Liège, this special sound happens because of unpredictable 'avalanche behaviour'.

The scientists call it 'avalanche behaviour' because the behaviour of the components is not independent, but tied to the behaviour of other components. For example, in a landslide, one tumbling pebble (a component) can set off other pebbles. In the same way, in a snow avalanche, one packet of snow (a component) can influence other packets of snow, and make a whole mountain side fall away.

Dr Vanderwalle looked at watery foams, which include champagne. He had difficulty in studying champagne bubbles, because they pop in about a thousandth of a second. So he worked with a water/soap mixture — in which the bubbles took much longer to pop. He soon realised that bubbles in the foam don't burst independently, instead, they affect each other. As one bubble pops, it emits a shock wave, which makes other bubbles pop, and so on.

Champagne noise is quite different from the 'white noise' (or static noise) you hear from a mis-tuned radio. White noise has a fairly constant loudness, but the loudness of the sound from champagne bubbles is not constant, but 'spiky'. It changes in loudness unpredictably. Champagne bubbles set off other bubbles in sympathy — creating little avalanches of sound. So next time you pour some champagne, hold the glass close to your ear, and listen to it get louder and softer.

champagne marketing

One reason for Professor Liger-Belair's study into champagne is to clarify the misconception most French people have that the smaller the bubbles, the better the champagne.

Champagne makers and champagne drinkers believe that 'bubbles' sell the champagne. They want lots of small bubbles, which keep generating for a long time. They also want a long-lived ring of foam, which should be very white with small bubbles.

For many years, champagne manufacturers didn't worry about competition from other sparkling white wines, because their champagne sold itself. But then cheap foreign sparkling white wines (at that time called champagnes) started cutting into the French champagne sales. So in the late 1980s, there was a huge increase in bubble research in France.

As a result of research and marketing, champagne is now more popular than ever, and sells about 270 million bottles each year.

when not to drink champagne

There are some places where you really shouldn't drink champagne, for example, where the pressure is very high.

One of the first tunnels under the Thames River in London was built using airlocks. The air in the tunnel had to be under enormous pressure to prevent water from flooding in. The airlocks kept the huge pressure in. Every day, the workers would enter and leave the tunnel through these airlocks.

Finally the two shafts from each side of the river met in the middle, and a celebratory dinner in the tunnel was held for the local politicians. To everybody's surprise, the champagne was flat. Even so, they drank the flat champagne. But when they emerged from the airlocks into the London night air, they were in for a surprise. The *'wine **popped** in their stomachs, distended their vests, all but frothed from their ears. One dignitary had to be rushed back into the depths to undergo champagne recompression.'*

Of course, with the benefit of hindsight, we now know the cause of the pain. The air pressure in the tunnel was so great, that it stopped the carbon dioxide in the champagne from coming out of its solution and making bubbles. When the dignitaries went through the airlocks back into the open air, the gas came out of the solution. Unfortunately this was inside their gut, causing immense pain, and one politician had to be rushed back underground — I guess this shows the incredible depths that alcohol forces some people to.

when to drink champagne

Madame Lily Bollinger, as in Bollinger Champagne, was one of the Grand Women of Champagne. These 'Champagne Widows' took over the family company after their husbands died, and indeed, made the company prosper. They included Madame Pommery and Madame Laurent-Perrier, as well as Madame Bollinger.

Madame Bollinger once famously said, *'I drink Champagne when I'm happy and when I'm sad. Sometimes I drink it when I'm alone. When I have company, I consider it obligatory. I trifle with it if I'm not hungry, and drink it when I am. Otherwise I never touch it — unless I'm thirsty.'*

Perhaps Napoleon Bonaparte spoke with real clarity when he said of champagne: *'In victory you deserve it, in defeat you need it.'*

The Right Champagne Glass

The champagne 'coup' was a short-stemmed saucer-shaped glass. One popular story claims that they were originally modelled on the breasts of Marie Antoinette, who was the Queen of France in the late 18th century. Another folk myth credits the breasts of Diane de Poitiers (1499–1566) who was the mistress of Henry II. In this version of the myth, Henry II especially loved her breasts, so she commissioned the glass blower at the Chateau d'Anet to make a replica of her breasts, as a present to Henry II.

The staff at Möet & Chandon in Reims tell a similar story, but with different people — Madame de Pompadur (1721–1764) and Louis XV. And Maurice des Ombiaux tells a similar story about Helen of Troy. I guess these stories just prove that male fantasies are totally irrational and illogical. You see, there are many problems with the coup.

First, the coup does not show off the beautiful and elegant bubble trains. Second, a coup has a wide rim, so the odours tend to dissipate. Third, a coup is so shallow and wide-brimmed, that it is unstable and can easily spill.

On the other hand, the long-stemmed flute, a slender glass with a tulip-shaped bowl, is much better, as it shows off the flow of bubbles when champagne is poured. The smaller surface area concentrates the aromas better as well. And because there's smaller surface area exposed to the air, it tends to preserve the chill of the champagne longer.

References

Ball, Philip, 'Bottoms Up', *Nature* (on-line), 1 March 2001.

Liger-Belair, Gérard *et al.*, 'On the velocity of expanding spherical gas bubbles rising in line in supersaturated hydroalcoholic solutions: application to bubble trains in carbonated beverages', *Langmuir*, vol. 16, no. 4, 2000, pp. 1889–1895.

Liger-Belair, Gérard, 'The science of bubbly', *Scientific American*, January 2003, pp. 68–73.

Perkowitz, Sidney, 'Gloriously bubbly', *New Scientist*, no. 2218, 25 December 1999/1 January 2000, pp. 58–61.

Plumb, Robert C., 'Chemical principles exemplified — champagne recompression', *Journal of Chemical Education*, vol. 48, no. 3, March 1971, pp. 154–155.

Vandewalle, N., Lentz, J.F., Dorbolo, S. and Brisbois, F., 'Avalanches of popping bubbles in collapsing foams', *Physical Review Letters*, vol. 86, no. 1, 1 January 2001, pp. 179–182.

Weiss, Peter, 'The physics of fizz', *Science News*, vol. 157, 6 May 2000, pp. 300–302.

thanks

In previous years I have always thanked all the people who have made each book possible. And I thank those people again because they are absolutely essential — Caroline Pegram, Adam Yazxhi and Sandra Davies (the editor known as Little Axe), who have worked under incredible pressure to get this book out.

However, I have never fully acknowledged my family. I have foolishly let all my 'writing time' eat into my 'family time'. And as I have written my way through 22 books, the writing time has gradually increased, and inversely, family time has decreased. No more.

My direct family has grown over the years. Mary, my beloved, has loved us, fed us, and guided us with her wisdom (and has made us laugh by being a lot of fun). Little Karl, whom I once photographed in a large cooking pot, is now taller than I am and will probably hit two metres before he stops growing. Little Alice has blossomed into a well-balanced teenager who knows her own mind. Little Lola has grown from a baby with purple feet into a five-year-old with immense curiosity. Brendan has grown from a boy who was thrilled to be an uncle at age five ('I'll be Uncle Five') to He-Of-The-Beautiful-Body who studies, surfs and is part of the family.

I am immensely proud of them all. And there are all the seemingly thousands of people in Mary's mega-family, who have given all of us a loving environment in which to swim.

Thanks, O Wonderful Family, and let's surf.

The Children's Hospital
Camperdown
NAME KARL KRUSZELNICKI
DESIGNATION R.M.O